Netty

源码剖析与应用

刘耀林 著

电子工业出版社
Publishing House of Electronics Industry
北京·BEIJING

内 容 简 介

Netty 涉及多线程技术、复杂数据结构与内存管理模型，它运用了各种设计模式及一些 TCP 的底层技术。本书对这些难点——进行攻破，让读者能快速掌握相关知识。

本书包含大量的分布式底层架构的编写，涉及多线程、负载均衡算法、性能调优、线上问题紧急处理等内容。本书通过非常简易的代码来讲解 Netty 在企业中的实际用法，通过对实例进行调试的方式对 Netty 源码进行了详细的剖析，力图使读者通过实际操作快速入门，并深入了解 Netty 底层的各个组件。

本书适合有一定 Java 基础的架构师、设计师、开发工程师、测试工程师，以及对 Java NIO 框架、Netty 感兴趣的相关人士阅读。

未经许可，不得以任何方式复制或抄袭本书之部分或全部内容。
版权所有，侵权必究。

图书在版编目（CIP）数据

Netty 源码剖析与应用 / 刘耀林著. —北京：电子工业出版社，2020.11
ISBN 978-7-121-39853-7

Ⅰ．①N… Ⅱ．①刘… Ⅲ．①JAVA 语言－程序设计Ⅳ．①TP312.8

中国版本图书馆 CIP 数据核字（2020）第 208000 号

责任编辑：宋亚东　　特约编辑：田学清
印　　刷：三河市良远印务有限公司
装　　订：三河市良远印务有限公司
出版发行：电子工业出版社
　　　　　北京市海淀区万寿路 173 信箱　　邮编：100036
开　　本：787×980　1/16　印张：17.5　字数：308 千字
版　　次：2020 年 11 月第 1 版
印　　次：2020 年 11 月第 1 次印刷
定　　价：89.00 元

凡所购买电子工业出版社图书有缺损问题，请向购买书店调换。若书店售缺，请与本社发行部联系，联系及邮购电话：(010) 88254888，88258888。
质量投诉请发邮件至 zlts@phei.com.cn，盗版侵权举报请发邮件至 dbqq@phei.com.cn。
本书咨询联系方式：(010) 51260888-819，faq@phei.com.cn。

前　言

Netty 是一款基于 NIO（非阻塞 I/O）开发的网络框架，与传统 BIO 相比，它的并发性能得到了很大的提高，而且更加节省资源。Netty 不仅封装了 NIO 操作的很多细节；在设计上还基于灵活、可扩展的事件驱动模型与高度可定制的线程模型，让 Netty 的应用更加灵活。作为一个被广泛使用的 Java 网络编程框架，Netty 在互联网领域、大数据分布式计算领域、游戏行业、物联网行业等都得到了广泛的应用。很多流行的大数据框架的核心通信模块也都使用的是 Netty，如 Elasticsearch、HBase、Flink 等。

为什么编写本书

如今，大数据已是互联网应用的大趋势，很多 Java 程序员都想转向大数据方向，而 Netty 不仅是大数据框架的核心，还是 Java 高并发中最火的框架之一。因此，学好 Netty 更有利于深入了解大数据框架底层源码。学习 Netty，不仅要学会如何运用它，还要对其底层原理、底层代码的编程技巧有深度的了解。目前，国内 Netty 方面的图书较少，尤其是既有深度，又比较简洁，并有企业级通用案例的图书更是凤毛麟角，这加大了很多 Java 程序员向大数据方向转变的难度，本书试图弥补这一空白。

本书主要基于 Netty 的稳定版本 Netty v4.1.38.Final（对以后的新版本也具有普适性）编写。本书的主要目的是帮助软件工程师读懂 Netty v4.1.38.Final 版本源码，并能开发一款类似 Dubbo 的分布式 RPC（Remote Procedure Call，远程过程调用），以及 Netty 的特性在 Flink 分布式流式计算框架中的实际应用。在本书中，作者对平常在编程中遇到的重点、难点进行了分析，并给予了充分的论述。对于一名软件工程师来说，本书可以使他们少走弯路，并更快地掌握 Netty v4.1.38.Final 版本源码及编程技巧。

关于本书作者

作者从 2012 年到 2017 年一直从事 Java 后台服务开发工作，在此期间曾创办大象在线分享网，网名夜行侠老师，录制 Netty 源码剖析教学视频，同时在多家互联网公司担任系统架构师，有丰富的 Java 工作实战经验。2018 年转型从事大数据研发工作，对 Flink、HBase、Kafka、Elasticsearch 等大数据组件进行了深入的研究。

本书主要内容

本书以实战为导向，深入解读 Netty 底层核心源码及架构设计。通过阅读本书，读者可以灵活运用 Netty 的特性、加深多线程实战应用、熟悉 Netty 的底层核心源码。全书共包含 2 个高级应用项目，分别为分布式 RPC 与 10 亿级任务调度和监控引擎。读者可以在实战过程中找到学习 Netty 的成就感，在学完 Netty 内存管理及 Netty 核心组件的源码剖析后，能够比较彻底地掌握 Netty。

如何阅读本书

在阅读本书前，读者应学过 Java 语言、网络通信等课程，并具备并发编程的能力。本书的章节安排是依据读者循序渐进地学习 Netty 的顺序设立的，建议初学者从前至后阅读。由于 Netty 源码复杂难懂，建议读者分以下几部分阅读本书。

- 第一部分：以实战为主，学会 Netty 客户端与服务端的长连接通信，灵活运用 Netty 的 Future 机制，同时对照第 3 章，尝试编写一套完整的分布式 RPC。

- 第二部分：仔细、反复地阅读 Netty 核心源码解读知识，主要包括 NioEventLoop 线程处理逻辑、Netty 的 Channel 功能设计、ByteBuf 缓冲区内存、内存泄漏检测机制。熟练掌握这部分内容有利于理解框架的整体实现原理。

- 第三部分：内存管理源码解读。这部分是 Netty 源码中最难掌握的，需要有一定的耐心。先了解 jemalloc 内存分配思想，再从底层 PoolChunk 的内存分配到上层的 PoolArena 对内存的整体管理。在深入学习具体的分配算法时，可把部分代码单独拿出来进行单元测试，以加深理解。

- 第四部分：Netty 的高级应用、线上问题分析和性能调优。这部分主要是作者的一些实战经验。通过前面对 Netty 的运用和源码解读，读者对 Netty 有了一定的了解，但还缺乏线上部署经验，以及高并发大数据的实际应用经验。这部分内容采用 Netty 时间轮实时监控 10 亿级任务，运用 Jmeter 长连接压测分布式 RPC，让读者在实际项目中更加全面和自信地使用 Netty。

源码阅读非常需要耐心，通过阅读 Netty 源码，读者会明显感觉到自身编程能力及源码阅读能力的提升，尤其是多线程编程能力。建议读者反复阅读至少 3 遍以上，直至对 Netty 的 Channel、Handler、NioEventLoop、ByteBuf 的各个方法都了如指掌。

致谢

我首先要特别感谢我的妻子谭小兰，写书需要大量的时间，我起初只是以视频在线教学的方式来总结自己的经验和收获，并分享给部分学员。是她在背后默默地支持我，我才能全身心投入书稿的写作去，从而与更多想深入学习 Netty 的读者分享本书。

同时感谢电子工业出版社博文视点宋亚东先生对本书的重视和他们为本书所做的一切。

由于水平有限，书中难免存在不足及疏漏，敬请专家和读者批评指正。

刘耀林

2020 年 11 月

读者服务

微信扫码回复：39853

- 获取作者提供的各种共享文档、线上直播、技术分享等免费资源。
- 加入本书读者交流群，与作者互动。
- 获取博文视点学院在线课程、电子书 20 元代金券。
- 获取本书配套的源代码。

轻松访问 http://www.broadview.com.cn/39853，并注册成为博文视点社区用户，可以在"下载资源"处下载本书源代码。您对书中内容的修改意见可在本书页面的"提交勘误"处提交，若被采纳，将获赠博文视点社区积分。在您购买电子书时，积分可用来抵扣相应金额。

目 录

第 1 章 Netty 基础篇 1

1.1 Netty 概述 1
1.2 Netty 服务端构建 2
1.3 Netty 客户端的运用 6
- 1.3.1 Java 多线程交互 6
- 1.3.2 Netty 客户端与服务端短连接 12
- 1.3.3 Netty 客户端与服务端长连接 18

1.4 小结 22

第 2 章 原理部分 23

2.1 多路复用器 23
- 2.1.1 NIO 与 BIO 的区别 24
- 2.1.2 epoll 模型与 select 模型的区别 25

2.2 Netty 线程模型 27
2.3 编码和解码 28
2.4 序列化 30
- 2.4.1 Protobuf 序列化 30

2.4.2 Kryo 序列化 31
2.5 零拷贝 33
2.6 背压 34
2.6.1 TCP 窗口 34
2.6.2 Flink 实时计算引擎的背压原理 36
2.7 小结 39

第 3 章 分布式 RPC 40
3.1 Netty 整合 Spring 41
3.2 采用 Netty 实现一套 RPC 框架 43
3.3 分布式 RPC 的构建 52
3.3.1 服务注册与发现 53
3.3.2 动态代理 68

第 4 章 Netty 核心组件源码剖析 81
4.1 NioEventLoopGroup 源码剖析 82
4.2 NioEventLoop 源码剖析 86
4.2.1 NioEventLoop 开启 Selector 87
4.2.2 NioEventLoop 的 run()方法解读 89
4.2.3 NioEventLoop 重新构建 Selector 和 Channel 的注册 98
4.3 Channel 源码剖析 100
4.3.1 AbstractChannel 源码剖析 101
4.3.2 AbstractNioChannel 源码剖析 102
4.3.3 AbstractNioByteChannel 源码剖析 107
4.3.4 AbstractNioMessageChannel 源码剖析 112

  4.3.5 NioSocketChannel 源码剖析 116

  4.3.6 NioServerSocketChannel 源码剖析 119

4.4 Netty 缓冲区 ByteBuf 源码剖析 120

  4.4.1 AbstractByteBuf 源码剖析 122

  4.4.2 AbstractReferenceCountedByteBuf 源码剖析 127

  4.4.3 ReferenceCountUpdater 源码剖析 129

  4.4.4 CompositeByteBuf 源码剖析 134

  4.4.5 PooledByteBuf 源码剖析 145

4.5 Netty 内存泄漏检测机制源码剖析 151

  4.5.1 内存泄漏检测原理 152

  4.5.2 内存泄漏器 ResourceLeakDetector 源码剖析 153

4.6 小结 164

第 5 章 Netty 读/写请求源码剖析 165

5.1 ServerBootstrap 启动过程剖析 165

5.2 Netty 对 I/O 就绪事件的处理 172

  5.2.1 NioEventLoop 就绪处理之 OP_ACCEPT 172

  5.2.2 NioEventLoop 就绪处理之 OP_READ（一）... 175

  5.2.3 NioEventLoop 就绪处理之 OP_READ（二）... 182

第 6 章 Netty 内存管理 195

6.1 Netty 内存管理策略介绍 195

6.2 PoolChunk 内存分配 197

  6.2.1 PoolChunk 分配大于或等于 8KB 的内存 197

  6.2.2 PoolChunk 分配小于 8KB 的内存 201

6.3　PoolSubpage 内存分配与释放 205

6.4　PoolArena 内存管理 214

6.5　RecvByteBufAllocator 内存分配计算 223

6.6　小结 227

第 7 章　Netty 时间轮高级应用 228

7.1　Netty 时间轮的解读 229

　　7.1.1　时间轮源码剖析之初始化构建 230

　　7.1.2　时间轮源码剖析之 Worker 启动线程 236

7.2　Netty 时间轮改造方案制订 239

7.3　时间轮高级应用之架构设计 241

7.4　时间轮高级应用之实战 10 亿级任务 243

7.5　小结 245

第 8 章　问题分析与性能调优 246

8.1　Netty 服务在 Linux 服务器上的部署 246

8.2　Netty 服务模拟秒杀压测 255

8.3　常见生产问题分析 264

8.4　性能调优 267

8.5　小结 270

Netty 基础篇

本章采用一段简洁的文字描述了 Netty 的作用。同时，通过一个长连接通信实例让读者可以快速掌握如何运用 Netty。

1.1 Netty 概述

普通开发人员在工作中一般很少接触 Netty，只有在阅读一些分布式框架底层源码时，才会发现底层通信模块大部分是 Netty，如 Dubbo、Flink、Spark、Elasticsearch、HBase 等流行的分布式框架。HBase 从 2.0 版本开始默认使用 Netty RPC Server，用 Netty 替代 HBase 原生的 RPC Server。至于微服务 Dubbo 和 RPC 框架（如 gRPC），它们的底层核心部分也都是 Netty。由此可见，不管是开发互联网 Java Web 后台，还是研发大数据，学好 Netty 都是很有必要的。

Netty 是一款流行的 Java NIO 框架，那么它有哪些特性呢？为什么其他优秀的 Java 框架的通信模块会选择 Netty 呢？使用过 Java NIO 的读者一定非常清楚，采用 NIO 编写一套高效且稳定的通信模块很不容易，没有一流的编程能力根本无法实现，并且无法做到在高并发情况下的可靠和高效。然而，Netty 这款优秀的开源框架却可以快速地开发高性能的面向协议的服务端和客户端。Netty 不仅易用、健壮、安全、高效，还可以轻松地自定义各种协议、采用各种序列化，并且它的可扩展性极强。

1.2 Netty 服务端构建

TCP 通信是面向连接的、可靠的、基于字节流的通信协议，有严格的客户端和服务端之分，本节运用 Netty 构建 TCP 服务端，同时为后面构建分布式 RPC 服务器做好准备。

Netty 服务端程序实现步骤如下。

（1）创建两个线程组，分别为 Boss 线程组和 Worker 线程组。Boss 线程专门用于接收来自客户端的连接；Worker 线程用于处理已经被 Boss 线程接收的连接。

（2）运用服务启动辅助类 ServerBootstrap 创建一个对象，并配置一系列启动参数，如参数 ChannelOption .SO_RCVBUF 和 ChannelOption .SO_SNDBUF 分别对应接收缓冲区和发送缓冲区的大小。

（3）当 Boss 线程把接收到的连接注册到 Worker 线程中后，需要交给连接初始化消息处理 Handler 链。由于不同的应用需要用到不同的 Handler 链，所以 Netty 提供了 ChannelInitializer 接口，由用户实现此接口，完成 Handler 链的初始化工作。

（4）编写业务处理 Handler 链，并实现对接收客户端消息的处理逻辑。

（5）绑定端口。由于端口绑定需要由 Boss 线程完成，所以主线程需要执行同步阻塞方法，等待 Boss 线程完成绑定操作。

在编码前先构建开发环境。开发工具选择 Spring Tool Suite 或 IntelliJ IDEA，其他工具选择 Maven v3.2.5、JDK v1.8、稳定版的 Netty v4.1.38.Final。首先打开 STS，构建一个 Maven 工程，并将 Netty 的依赖放入 pom.xml 中。具体实现代码如下：

```xml
<dependency>
<groupId>io.netty</groupId>
<artifactId>netty-all</artifactId>
<version>4.1.38.Final</version>
</dependency>
```

Netty 的 Maven 工程构建图如图 1-1 所示。

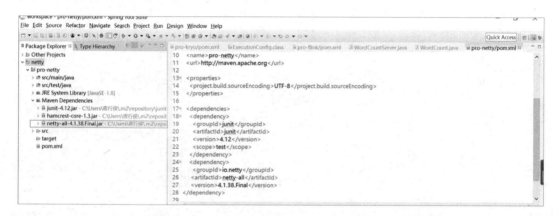

图 1-1　Netty 的 Maven 工程构建图

按照服务端程序实现步骤新建一个 Netty 服务类，具体代码如下：

```java
package com.itjoin.pro_netty.server;
import io.netty.bootstrap.ServerBootstrap;
import io.netty.channel.ChannelFuture;
import io.netty.channel.ChannelInitializer;
import io.netty.channel.ChannelOption;
import io.netty.channel.EventLoopGroup;
import io.netty.channel.nio.NioEventLoopGroup;
import io.netty.channel.socket.SocketChannel;
import io.netty.channel.socket.nio.NioServerSocketChannel;
public class Netty服务 {
    public static void main(String[] args) throws Exception {
        /**
```

```
 * 新建两个线程组，Boss 线程组启动一条线程，监听 OP_ACCEPT 事件
 * Worker 线程组默认启动 CPU 核数*2 的线程
 * 监听客户端连接的 OP_READ 和 OP_WRITE 事件，处理 I/O 事件
 */
EventLoopGroup bossGroup = new NioEventLoopGroup(1);
EventLoopGroup workerGroup = new NioEventLoopGroup();
try {
    //ServerBootstrap 为 Netty 服务启动辅助类
    ServerBootstrap serverBootstrap = new ServerBootstrap();
    serverBootstrap.group(bossGroup, workerGroup);
    //设置 TCP Socket 通道为 NioServerSocketChannel
    //若是 UDP 通信，则将其设置为 DatagramChannel
    serverBootstrap.channel(NioServerSocketChannel.class);
    //设置一些 TCP 参数
    serverBootstrap.option(ChannelOption.SO_BACKLOG,128)
    /**
     * 当有客户端链路注册读写事件时，初始化 Handler
     * 并将 Handler 加入管道中
     */
    .childHandler(new ChannelInitializer<SocketChannel>() {
    @Override
    protected void initChannel(SocketChannel ch) throws Exception {
        /**
         * 向 Worker 线程的管道双向链表中添加处理类 ServerHandler
         * 整个处理流向如下：HeadContext-channelRead 读数据-->ServerHandler-channelRead
         * 读取数据进行业务逻辑判断，最后将结果返回给客户端-->TailContext-write->
         * HeadContext-write
         */
        ch.pipeline().addLast(new ServerHandler());}});
    //同步绑定端口
    ChannelFuture future = serverBootstrap.bind(8080).sync();
    //阻塞主线程，直到 Socket 通道被关闭
    future.channel().closeFuture().sync();
} catch (Exception e) {
    e.printStackTrace();
}finally {
    //最终关闭线程组
```

```
            workerGroup.shutdownGracefully();
            bossGroup.shutdownGracefully();
        }
    }
}
```

服务端还需要编写一个业务逻辑处理 Handler（名称为 ServerHandler），这个 Handler 需要读取客户端数据，并对请求进行业务逻辑处理，最终把响应结果返回给客户端。ServerHandler 需要继承 ChannelInboundHandlerAdapter，它是 ChannelInboundHandler 的子类，这跟 Netty 的处理数据流向有关。当 NioEventLoop 线程从 Channel 读取数据时，执行绑定在 Channel 的 ChannelInboundHandler 对象上，并执行其 channelRead() 方法。具体实现代码如下：

```
package com.itjoin.pro_netty.server;
import java.nio.charset.Charset;
import io.netty.buffer.ByteBuf;
import io.netty.channel.ChannelHandlerContext;
import io.netty.channel.ChannelInboundHandlerAdapter;
//@Sharable 注解表示此 Handler 对所有 Channel 共享，无状态，注意多线程并发
@ChannelHandler.Sharable
public class ServerHandler extends ChannelInboundHandlerAdapter {

    /**
     * 读取客户端发送的数据
     */
    @Override
    public void channelRead(ChannelHandlerContext ctx, Object msg) {
        if(msg instanceof ByteBuf) {
            //ByteBuf 的 toString()方法把二进制数据转换成字符串，默认编码 UTF-8
            System.out.println(((ByteBuf)msg).toString(
                Charset.defaultCharset()));
        }
        ctx.channel().writeAndFlush("msg has recived!");
    }
}
```

1.3　Netty 客户端的运用

Netty 除了可以编写高性能服务端，还有配套的非阻塞 I/O 客户端，关于客户端与服务端的通信，涉及多线程数据交互，并运用了 JDK 的锁和多线程。

1.3.1　Java 多线程交互

本小节编写了一个 Java 多线程实例，为后续介绍 Netty 客户端做铺垫。此实例模拟 1 条主线程循环写数据，另外 99 条子线程，每条子线程模拟睡眠 1s，再给主线程发送结果，最终主线程阻塞获取 99 条子线程响应的结果数据。

此实例有以下几个类。

第一个类 FutureMain：主线程，拥有启动 main() 方法和实例的总体处理逻辑（请求对象的生成、子线程的构建和启动、获取子线程的响应结果）。

第二个类 RequestFuture：模拟客户端请求类，主要用于构建请求对象（拥有每次的请求 id，并对每次请求对象进行了缓存），最核心的部分在于它的同步等待和结果通知方法。

第三个类 SubThread：子线程，用于模拟服务端处理，根据主线程传送的请求对象 RequestFuture 构建响应结果，等待 1s 后，调用 RequestFuture 的响应结果通知方法将结果交互给主线程。

第四个类 Response：响应结果类，拥有响应 id 和结果内容。只有响应 id 和请求 id 一致，才能从请求缓存中获取当前响应结果的请求对象。

如图 1-2 为多线程交互数据 UML 图。

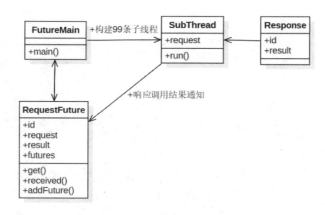

图 1-2 多线程交互数据 UML 图

具体实现代码如下。

（1）FutureMain 类：

```
package com.itjoin.pro_netty.asyn;
import java.util.ArrayList;
import java.util.List;
/**
 * 主线程类，主线程模拟发送请求，开启额外线程模拟获取响应结果
 * 并异步地将响应结果返回主线程
 */
public class FutureMain {
    public static void main(String[] args) {
        //请求列表
        List<RequestFuture> reqs = new ArrayList<>();
        /**
         * 此处用 for 循环模拟连续发送 100 个请求
         * 异步构建 100 条线程获取请求，并返回响应结果
         * 当然，此处还可以用线程池模拟构建 100 条线程发送请求
         * 然后主线程等待所有子线程获取对应的响应结果。希望读者能对代码进行相应的修改
         */
        for(int i=1;i<100;i++) {
            //请求 id
            long id = i;
```

```java
        //构建请求对象
        RequestFuture req = new RequestFuture();
        req.setId(id);
        //设置请求内容
        req.setRequest("hello world");
        //把请求缓存起来
        RequestFuture.addFuture(req);
        //把请求加入请求列表
        reqs.add(req);
        //模拟发送请求
        sendMsg(req);
        //模拟线程，获取对应的请求
        SubThread subThread = new SubThread(req);
        subThread.start();
    }
    for(RequestFuture req : reqs) {
        //主线程获取响应结果
        Object result = req.get();
        //输出结果
        System.out.println(result.toString());
    }
}
private static void sendMsg(RequestFuture req) {
    System.out.println("客户端发送数据,请求id为====="+req.getId());
}
}
```

（2）RequestFuture 类：

```java
package com.itjoin.pro_netty.asyn;
import java.util.Map;
import java.util.concurrent.ConcurrentHashMap;
public class RequestFuture {
    //请求缓存类,key 为每次请求 id, value 为请求对象
    public static Map<Long,RequestFuture> futures
        = new ConcurrentHashMap<Long,RequestFuture>();
    //对于每次请求 id,可以设置原子性增长
    private long id;
```

```java
//请求参数
private Object request;
//响应结果
private Object result;
//超时时间默认为5s
private long timeout=5000;
//把请求放入缓存中
public static void addFuture(RequestFuture future) {
    futures.put(future.getId(), future);
}
/**同步获取响应结果*/
public Object get() {
    /**此处可以把同步块与wait换成ReentrantLock与Condition*/
    synchronized (this) {
        while(this.result==null) {
            try {
                /**主线程默认等待5s,然后查看是否获取到结果*/
                this.wait(timeout);
            } catch (InterruptedException e) {
                e.printStackTrace();
            }
        }

    }
    return this.result;
}
/**异步线程将结果返回主线程*/
public static void received(Response resp) {
    RequestFuture future = futures.remove(resp.getId());
    //设置响应结果
    if(future!=null)
        future.setResult(resp.getResult());
    /**通知主线程*/
    synchronized (future) {
        future.notify();
    }
}
```

```java
    public long getId() {
        return id;
    }
    public void setId(long id) {
        this.id = id;
    }
    public Object getRequest() {
        return request;
    }
    public void setRequest(Object request) {
        this.request = request;
    }
    public Object getResult() {
        return result;
    }
    public void setResult(Object result) {
        this.result = result;
    }
}
```

（3）SubThread 类：

```java
package com.itjoin.pro_netty.asyn;
/**
 * 子线程，此线程模拟 Netty 的异步响应结果
 */
public class SubThread extends Thread{
    private RequestFuture request;
    public SubThread(RequestFuture request) {
        this.request = request;
    }
    @Override
    public void run() {
        //模拟额外线程获取响应结果
        Response resp = new Response();
        /**
         * 此处 id 为请求 id，模拟服务器接收到请求后
         * 将请求 id 直接赋给响应对象 id
```

```
    */
    resp.setId(request.getId());
    //为响应结果赋值
    resp.setResult("server response"+Thread.currentThread().getId());
    //子线程模拟睡眠1s
    try {
        Thread.sleep(1000);
    } catch (InterruptedException e) {
        e.printStackTrace();
    }
    //此处将响应结果返回主线程
    RequestFuture.received(resp);
  }
}
```

（4）Response 类：

```
package com.itjoin.pro_netty.asyn;
public class Response {
    private long id;
    private Object result;
    public long getId() {
        return id;
    }
    public void setId(long id) {
        this.id = id;
    }
    public Object getResult() {
        return result;
    }
    public void setResult(Object result) {
        this.result = result;
    }
}
```

运行 FutureMain 的 main()方法，控制台会打印出子线程 SubThread 返回的 Response 消息。感兴趣的读者可通过以下两个问题对代码进行相应的修改。

（1）使用 ReentrantLock 与 Condition 更换同步关键词，以及 Object 类的 notify()和 wait()方法。

（2）将主线程 for 循环构建请求而阻塞同步发送修改成线程池 Executors.newFixedThreadPool 生成 100 条线程异步请求。

1.3.2　Netty 客户端与服务端短连接

Netty 是一个异步网络处理框架，使用了大量的 Future 机制，并在 Java 自带 Future 的基础上增加了 Promise 机制，从而使异步编程更加方便简单。本小节采用 Netty 客户端与服务端实现短连接异步通信的方式，加深读者对多线程的灵活运用的认识，并帮助读者初步了解 Netty 客户端。

Netty 客户端的线程模型比服务端的线程模型简单一些，它只需一个线程组，底层采用 Java 的 NIO，通过 IP 和端口连接目标服务器，请求发送和接收响应结果数据与 Netty 服务端编程一样，同样需要经过一系列 Handler。请求发送从 TailContext 到编码器 Handler，再到 HeadContext；接收响应路径从 HeadContext 到解码器 Handler，再到业务逻辑 Handler（此处提到的数据流的编码和解码会在第 5 章进行详细讲解）。TCP 网络传输的是二进制数据流，且会源源不断地流到目标机器上，如果没有对数据流进行一些额外的加工处理，那么将无法区分每次请求的数据包。编码是指在传输数据前，对数据包进行加工处理，解码发生在读取数据包时，根据加工好的数据的特点，解析出正确的数据包。客户端与服务端的交互流程如图 1-3 所示。

图 1-3 中共有 8 个处理流程，分别如下。

① Netty 客户端通过 IP 和端口连接服务端，并准备好 JSON 数据包。

② JSON 数据包发送到网络之前需要经过一系列编码器，并最终被写入 Socket 中，发送给 Netty 服务端。

③ Netty 服务端接收到 Netty 客户端发送的数据流后，先经过一系列解码器，把客户端发

送的 JSON 数据包解码出来，然后传递给 ServerHandler 实例。

④ ServerHandler 实例模拟业务逻辑处理。

⑤ Netty 服务端把处理后的结果返回 Netty 客户端。

⑥ Netty 服务端同样需要经过一系列编码器，最终将响应结果发送到网络中。

⑦ Netty 客户端解码器接收响应结果字节流并对其进行解码，然后把响应的 JSON 结果数据返回给 ClientHandler。

⑧ 由于 ClientHandler 运行在 NioEventLoop 线程上，所以结果数据在返回主线程时需要用到 Netty 的 Promise 机制，以实现多线程数据交互。

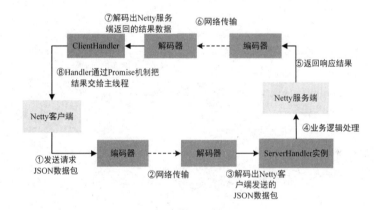

图 1-3　客户端与服务端的交互流程

Netty 服务端的 Netty 服务类在 1.2 节的基础上新增了长度编码器和解码器，具体代码如下：

```
serverBootstrap.option(ChannelOption.SO_BACKLOG, 128)
        /** 当有客户端链路注册读写事件时，初始化 Handler，并把 Handler 加入管道中*/
        .childHandler(new ChannelInitializer<SocketChannel>() {
@Override
protected void initChannel(SocketChannel ch) throws Exception {
// 新增，以前缀为 4B 的 int 类型作为长度的解码器
// 第一个参数是包的最大长度；第二个参数是长度值偏移量，由于编码时长度值在最前面
// 无偏移，所以此处设置为 0
// 第三个参数是长度值占用的字节数
```

```
// 第四个参数是长度值的调节, 假设请求包的大小是 20B
// 若长度值不包含本身则应该是 20B, 若长度值包含本身则应该是 24B, 需要调整 4 个字节
// 第五个参数是在解析时需要跳过的字节数 (此处为 4)
ch.pipeline().addLast(new LengthFieldBasedFrameDecoder(Integer.MAX_VALUE,
                                                        0, 4, 0, 4));
// 把接收到的 ByteBuf 数据包转换成 String
ch.pipeline().addLast(new StringDecoder());
/**
 * 向 Worker 线程的管道双向链表中添加处理类 ServerHandler
 * 整个处理流向如下: * HeadContext-channelRead 读数据-->
 * LengthFieldBasedFrameDecoder-->StringDecoder
 * -->ServerHandler-channelRead 读取数据进行业务逻辑判断
 * 最后将结果返回客户端-->TailContext-write->StringEncoder-->
 * LengthFieldPrepender -->HeadContext-write
 */
ch.pipeline().addLast(new ServerHandler());
// 在消息体前面新增 4 个字节的长度值, 第一个参数是长度值占用的字节数
// 第二个参数是长度值的调节, 表明是否包含长度值本身
ch.pipeline().addLast(new LengthFieldPrepender(4, false));
// 把字符串消息转换成 ByteBuf
ch.pipeline().addLast(new StringEncoder());
// 注意解码器和编码器的顺序
// 两者的执行顺序正好相反, 解码器执行顺序从上往下, 编码器执行顺序从下往上
    }
});
```

服务端业务逻辑处理类 ServerHandler 的 channelRead()方法需要返回 Response 对象, 同时获取请求 id, 具体变动代码如下:

```
public void channelRead(ChannelHandlerContext ctx, Object msg) {
    // if(msg instanceof ByteBuf) {
    //ByteBuf 的 toString()方法把二进制数据转换成字符串, 默认编码为 UTF-8
    //System.out.println(((ByteBuf)msg).toString
    //(Charset.defaultCharset()));
    //   }
    //ctx.channel().writeAndFlush("msg has recived!");
    //获取客户端发送的请求, 并将其转换成 RequestFuture 对象
    //由于经过了 StringDecoder 解码器, 所以 msg 为 String 类型
```

```java
        RequestFuture request
                = JSONObject.parseObject(msg.toString(),RequestFuture.class);
    //获取请求id
    long id = request.getId();
    System.out.println("请求信息为==="+msg.toString());
    //构建响应结果
    Response response = new Response();
    response.setId(id);
    response.setResult("服务器响应ok");
    //把响应结果返回客户端
    ctx.channel().writeAndFlush(JSONObject.toJSONString(response));
}
```

客户端启动类 NettyClient 和 Netty 服务相似,辅助类用 Bootstrap 替代 ServerBootstrap,同时引入了异步处理类 DefaultPromise,用于异步获取服务端的响应结果,具体实现代码如下:

```java
package com.itjoin.pro_netty.client;
import java.util.concurrent.ExecutionException;
import com.alibaba.fastjson.JSONObject;
import com.itjoin.pro_netty.asyn.RequestFuture;
import com.itjoin.pro_netty.asyn.Response;
import io.netty.bootstrap.Bootstrap;
import io.netty.buffer.PooledByteBufAllocator;
import io.netty.channel.ChannelFuture;
import io.netty.channel.ChannelInitializer;
import io.netty.channel.ChannelOption;
import io.netty.channel.EventLoopGroup;
import io.netty.channel.nio.NioEventLoopGroup;
import io.netty.channel.socket.nio.NioSocketChannel;
import io.netty.handler.codec.LengthFieldBasedFrameDecoder;
import io.netty.handler.codec.LengthFieldPrepender;
import io.netty.handler.codec.string.StringDecoder;
import io.netty.handler.codec.string.StringEncoder;
import io.netty.util.concurrent.DefaultPromise;
import io.netty.util.concurrent.Promise;
public class NettyClient {
```

```java
public static EventLoopGroup group=null;
public static Bootstrap boostrap=null;
static{
    //客户端启动辅助类
    boostrap = new Bootstrap();
    //开启一个线程组
    group = new NioEventLoopGroup();
    //设置Socket通道
    boostrap.channel(NioSocketChannel.class);
    boostrap.group(group);
    //设置内存分配器
    boostrap.option(ChannelOption.ALLOCATOR,PooledByteBufAllocator.DEFAULT);
}
public static void main(String[] args) throws ExecutionException {
    try {
        //新建一个promise对象
        Promise<Response> promise = new DefaultPromise<>(group.next());
        //业务Handler
        final ClientHandler handler = new ClientHandler();
        //把promise对象赋给handler,用于获取返回服务端的响应结果
        handler.setPromise(promise);
        //把handler对象加入管道中
        boostrap.handler(new ChannelInitializer<NioSocketChannel>() {
            @Override
            protected void initChannel(NioSocketChannel ch)
                throws Exception {
                ch.pipeline().addLast(
                    new LengthFieldBasedFrameDecoder(Integer.MAX_VALUE,
                        0, 4, 0, 4));
                //把接收到的ByteBuf数据包转换成String
                ch.pipeline().addLast(new StringDecoder());
                //业务逻辑处理Handler
                ch.pipeline().addLast(handler);
                ch.pipeline().addLast(new LengthFieldPrepender(4, false));
                //把字符串消息转换成ByteBuf
                ch.pipeline().addLast(new
                    StringEncoder(Charset.forName("utf-8")));
```

```
        }
    });
    //连接服务器
    ChannelFuture future = boostrap.connect("127.0.0.1", 8080).sync();
    //构建 request 请求
    RequestFuture request = new RequestFuture();
    //设置请求 id，此处请求 id 可以设置为自动自增模式
    //可以采用 AtomicLong 类的 incrementAndGet()方法
    request.setId(1);
    //请求消息内容，此处内容可以是任意的 Java 对象
    request.setRequest("hello world!");
    //转换成 JSON 格式发送给编码器 StringEncode，
    //StringEncode 编码器再发送给 LengthFieldPrepender 长度编码器，最终写到
    //TCP 缓存中，并传送给客户端
    String requestStr = JSONObject.toJSONString(request);
    future.channel().writeAndFlush(requestStr);
    //同步阻塞等待响应结果
    Response response = promise.get();
    //打印最终结果
    System.out.println(JSONObject.toJSONString(response));
} catch (InterruptedException e) {
    e.printStackTrace();
}
    }
}
```

客户端业务逻辑处理 ClientHandler，接收服务端的响应数据，并运用 Promise 唤醒主线程，实现代码如下：

```
package com.itjoin.pro_netty.client;
import com.alibaba.fastjson.JSONObject;
import com.itjoin.pro_netty.asyn.Response;
import io.netty.channel.ChannelHandlerContext;
import io.netty.channel.ChannelInboundHandlerAdapter;
import io.netty.util.concurrent.Promise;
public class ClientHandler extends ChannelInboundHandlerAdapter {
    private Promise<Response> promise;
    @Override
```

```java
public void channelRead(ChannelHandlerContext ctx, Object msg)
    throws Exception {
  //读取服务端返回的响应结果,并将其转换成 Response 对象
  //由于经过了 StringDecoder 解码器,所以 msg 为 String 类型
  Response response
      = JSONObject.parseObject(msg.toString(),Response.class);
  //设置响应结果并唤醒主线程
  promise.setSuccess(response);
}
public Promise<Response> getPromise() {
  return promise;
}
public void setPromise(Promise<Response> promise) {
  this.promise = promise;
}
```

通过以上的介绍,对 Netty 客户端会有一个基本的了解,能正常发送与接收数据。上述用法看起来没什么问题,但性能偏弱。因为在每次发送请求时都需要创建连接,还不如直接使用普通 Socket 作为客户端。可以运用连接池优化解决,把短连接先放入连接池,然后从连接池中获取每次请求的连接,感兴趣的读者可以采用 Apache Commons Pool 重构上述代码。

1.3.3　Netty 客户端与服务端长连接

本小节使用长连接解决 1.3.2 小节中的性能问题,通常各公司内部的 RPC 通信一般会选择长连接通信模式。在改造代码前,请先思考以下两个问题。

(1)改造成长连接后,ClientHandler 不能每次都在 main()方法中构建,promise 对象无法通过主线程传送给 ClientHandler,那么此时主线程如何获取 NioEventLoop 线程的数据呢?

(2)主线程每次获取的响应结果对应的是哪次请求呢?

- 通过多线程交互数据实例的学习,很显然第一个问题可以通过多线程数据交互来解决。首先对 Netty 客户端创建的连接进行静态化处理,以免每次调用时都需要重复创建;然后在给服务端发送请求后运用 RequestFuture 的 get()方法同步等待获取响应结果,以替代

Netty 的 Promise 同步等待;最后用 RequestFuture.received 替代 ClientHandler 的 Promise 异步通知。

- 第二个问题的解决方案:每次请求带上自增唯一的 id,客户端需要把每次请求先缓存起来,同时服务端在接收到请求后,会把请求 id 放入响应结果中一起返回客户端。

NettyClient 的连接需要进行静态化处理,同时,改成 RequestFutrue 的 get()方法获取异步响应结果。具体的改造后的 NettyClient 代码如下:

```
package com.itjoin.pro_netty.client;
import java.util.concurrent.ExecutionException;
import com.alibaba.fastjson.JSONObject;
import com.itjoin.pro_netty.asyn.RequestFuture;
import io.netty.bootstrap.Bootstrap;
import io.netty.buffer.PooledByteBufAllocator;
import io.netty.channel.ChannelFuture;
import io.netty.channel.ChannelInitializer;
import io.netty.channel.ChannelOption;
import io.netty.channel.EventLoopGroup;
import io.netty.channel.nio.NioEventLoopGroup;
import io.netty.channel.socket.nio.NioSocketChannel;
import io.netty.handler.codec.LengthFieldBasedFrameDecoder;
import io.netty.handler.codec.LengthFieldPrepender;
import io.netty.handler.codec.string.StringDecoder;
import io.netty.handler.codec.string.StringEncoder;
public class NettyClient {
    public static EventLoopGroup group=null;
    public static Bootstrap boostrap=null;
    public static ChannelFuture future=null;
    static{
        //客户端启动辅助类
        boostrap = new Bootstrap();
        //开启一个线程组
        group = new NioEventLoopGroup();
        //设置Socket通道
        boostrap.channel(NioSocketChannel.class);
        boostrap.group(group);
```

```java
//设置内存分配器
boostrap.option(ChannelOption.ALLOCATOR, PooledByteBufAllocator.DEFAULT);
final ClientHandler handler = new ClientHandler();
//把handler加入管道中
boostrap .handler(new ChannelInitializer<NioSocketChannel>() {
        @Override
        protected void initChannel(NioSocketChannel ch)
                throws Exception {
   ch.pipeline().addLast(new LengthFieldBasedFrameDecoder(Integer.MAX_VALUE,
                     0, 4, 0, 4));
   //把接收到的ByteBuf数据包转换成String
   ch.pipeline().addLast(new StringDecoder());
   //业务逻辑处理Handler
   ch.pipeline().addLast(handler);
   ch.pipeline().addLast(new LengthFieldPrepender(4, false));
   //把字符串消息转换成ByteBuf
   ch.pipeline().addLast(new StringEncoder(Charset.forName("utf-8")));
}
});
      //连接服务器
      try {
          future = boostrap.connect("127.0.0.1", 8080).sync();
      } catch (InterruptedException e) {
          e.printStackTrace();
      }
   }
   public Object sendRequest(Object msg) {
      try {
         //构建request
         RequestFuture request = new RequestFuture();
         //请求消息内容,此处内容可以是任意的Java对象
         request.setRequest(msg);
         //将请求消息转换成JSON并发送给编码器StringEncode
         //StringEncode编码器再发送给LengthFieldPrepender长度编码器
         //最终写到TCP缓存中并传送给客户端
          String requestStr = JSONObject.toJSONString(request);
         //注意:在进行单个连接时,future可以静态化,但在多个连接的情况下,future不可以静态化
```

```
        future.channel().writeAndFlush(requestStr);
        //同步等待响应结果,只有当promise被赋值后,程序才会继续向下执行
        Object result = request.get();
        return result;
    } catch (Exception e) {
        e.printStackTrace();
        throw e;
    }
}
public static void main(String[] args) throws ExecutionException {
    NettyClient client = new NettyClient();
    for(int i=0;i<100;i++) {
        Object result = client.sendRequest("hello");
        System.out.println(result);
    }
}
}
```

下面对 RequestFuture 类也进行一些改动,新增了一个 id 自增属性 AtomicLong aid,在构造方法上,将自增 aid 赋给请求 id,同时把当前请求对象加入全局缓存 futures 中,代码如下:

```
//自增id
private static final AtomicLong aid=new AtomicLong(1);
public RequestFuture() {
    //当前值新增1并将结果返回给id
    id = aid.incrementAndGet();
    //在构建请求时,需要把请求加入缓存中
    addFuture(this);
}
```

将 ClientHandler 的 channelRead()方法中的 promise 改为 RequestFuture.received,代码如下:

```
public void channelRead(ChannelHandlerContext ctx, Object msg)
        throws Exception {
    //读取服务器返回的响应结果,并将其转换成 Response 对象
    //由于经过了 StringDecoder 解码器,所以 msg 为 String 类型
    Response response
        = JSONObject.parseObject(msg.toString(),Response.class);
    RequestFuture.received(response);
```

```
        //设置响应结果并唤醒主线程
        //promise.setSuccess(response);
}
```

服务端 ServerHandler 类将返回的结果加上对应的请求 id，客户端与服务端长连接响应输出如图 1-4 所示。通过图 1-4 发现，服务器响应的结果有序地输出在客户端控制台上。不妨思考，要如何修改代码，才能让客户端输出的结果是无序的呢？

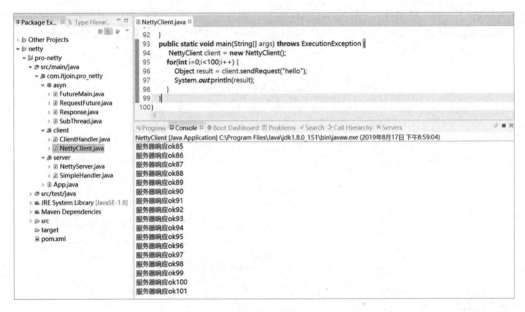

图 1-4　客户端与服务端长连接响应输出

1.4　小结

本章虽然是 Netty 的基础应用部分，但是涉及 Java 多线程交互、Netty 客户端与服务端的长连接通信，为编写分布式 RPC 打好了基础。在进行后续的 Netty 多线程编程时，会遇到各种问题。例如，从表面上看，Netty 客户端只用一条线程就能完成与服务端的数据交互，为何要使用线程组？然而在实际应用中，Netty 服务会部署在多台机器上，而客户端与服务端的连接也会有多条，这些连接链路 Channel 可以注册在同一个 Worker 线程组中。在学习 Netty 时，要多发现问题并多思考，以找到一个让自己满意的答案。

原理部分

2.1 多路复用器

通过第 1 章的介绍，我们对 Netty 有了初步的了解，但距离完全掌握 Netty，并编写高性能的 RPC 服务还有一定的差距。本章主要对 Netty 和 NIO 的一些特性进行梳理，以了解它们的底层原理，为后面编写 RPC 分布式服务打好理论基础。

NIO 有一个非常重要的组件——多路复用器，其底层有 3 种经典模型，分别是 epoll、select 和 poll。与传统 I/O 相比，一个多路复用器可以处理多个 Socket 连接，而传统 I/O 对每个 Socket 连接都需要一条线程去同步阻塞处理。NIO 有了多路复用器后，只需一条线程即可管理多个 Socket 连接的接入和读写事件。Netty 的多路复用器默认调用的模型是 epoll 模型。它除了 JDK

自带的 epoll 模型的封装，还额外封装了一套，它们都是 epoll 模型的封装，只是 JDK 的 epoll 模型是水平触发的，而 Netty 采用 JNI 重写的是边缘触发。

2.1.1　NIO 与 BIO 的区别

NIO 为同步非阻塞 I/O，BIO 为同步阻塞 I/O。阻塞 I/O 与非阻塞 I/O 的区别如下。

阻塞 I/O：例如，客户端向服务器发送 10B 的数据，若服务器一次只接收到 8B 的数据，则必须一直等待后面 2B 大小的数据的到来，在后面 2B 的数据未到达时，当前线程会阻塞在接收函数处。

非阻塞 I/O：从 TCP 缓冲区中读取数据，缓冲区中的数据可分多次读取，不会阻塞线程。例如，已知后面将有 10B 的数据发送过来，但是如果现在缓冲区只收到 8B 的数据，那么当前线程就会读取这 8B 的数据，读完后立即返回，等另外 2B 的数据发来时再去读取。NIO 中的多路复用器管理成千上万条 Socket 连接。当多路复用器每次从 TCP 缓冲区读数据时，若有些客户端数据包未能全部到达，且读取数据的线程是在阻塞的情况下，则只有全部数据到达时才能返回，这样不仅性能弱，还不可控，无法预测等待的时间。

图 2-1 为 BIO 服务器流程图，图 2-2 为 NIO 服务器整体流程图。

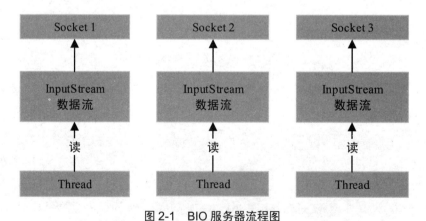

图 2-1　BIO 服务器流程图

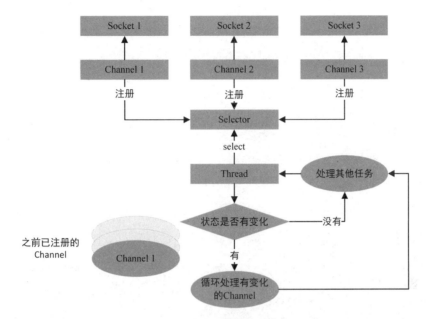

图 2-2　NIO 服务器整体流程图

从图 2-1 和图 2-2 中可以明显地看出，NIO 比 BIO 复杂得多，NIO 主要是多了 Selector。Selector 能监听多个 Channel，当运行 select()方法时，会循环检测是否有就绪事件的 Channel。只需一条线程即可管理多个 Channel，而且对 Channel 的读/写采用的都是非阻塞 I/O。与 BIO 相比，NIO 同时接入的 Channel 会更多、资源利用率会更高。因为 BIO 的一条线程只能对单个 Channel 进行阻塞读/写，处理完后才能继续接入并处理其他 Channel，并发处理能力太弱。

2.1.2　epoll 模型与 select 模型的区别

I/O 多路复用器单个进程可以同时处理多个描述符的 I/O，Java 应用程序通过调用多路复用器来获取有事件发生的文件描述符，以进行 I/O 的读/写操作。多路复用器常见的底层实现模型有 epoll 模型和 select 模型，本节详细介绍它们各自的特点。

select 模型有以下 3 个特点。

（1）select 模型只有一个 select 函数，每次在调用 select 函数时，都需要把整个文件描述符集合从用户态拷贝到内核态，当文件描述符很多时，开销会比较大。

（2）每次在调用 select 函数时，内核都需要遍历所有的文件描述符，这个开销也很大，尤其是当很多文件描述符根本就无状态改变时，也需要遍历，浪费性能。

（3）select 可支持的文件描述符有上限，可监控的文件描述符个数取决于 sizeOf(fd_set) 的值。如果 sizeOf(fd_set)=512，那么此服务器最多支持 512×8=4096 个文件描述符。

epoll 模型比 select 模型复杂，epoll 模型有三个函数。第一个函数为 int epoll_create(int size)，用于创建一个 epoll 句柄。第二个函数为 int epoll_ctl(int epfd,int op,int fd,struct epoll_event *event)，其中，第一个参数为 epoll_create 函数调用返回的值；第二个参数表示操作动作，由三个宏（EPOLL_CTL_ADD 表示注册新的文件描述符到此 epfd 上，EPOLL_CTL_MOD 表示修改已经注册的文件描述符的监听事件，EPOLL_CTL_DEL 表示从 epfd 中删除一个文件描述符）来表示；第三个参数为需要监听的文件描述符；第四个参数表示要监听的事件类型，事件类型也是几个宏的集合，主要是文件描述符可读、可写、发生错误、被挂断和触发模式设置等。epoll 模型的第三个函数为 epoll_wait，表示等待文件描述符就绪。epoll 模型与 select 模型相比，在以下这些地方进行了改善。

- 所有需要监听的文件描述符只需在调用第二个函数 int epoll_ctl 时拷贝一次即可，当文件描述符状态发生改变时，内核会把文件描述符放入一个就绪队列中，通过调用 epoll_wait 函数获取就绪的文件描述符。
- 每次调用 epoll_wait 函数只会遍历状态发生改变的文件描述符，无须全部遍历，降低了操作的时间复杂度。
- 没有文件描述符个数的限制。
- 采用了内存映射机制，内核直接将就绪队列通过 MMAP 的方式映射到用户态，避免了内存拷贝带来的额外性能开销。

了解这两种多路复用器模型的特点主要是为了加深对 NIO 底层原理的理解，同时对深入了解 Netty 源码有很大的帮助。

2.2 Netty 线程模型

从本节开始了解 Netty 的特性,首先从线程模型开始。在第 1 章中,讲到过两个线程组,即 Boss 线程组和 Worker 线程组。其中,Boss 线程组一般只开启一条线程,除非一个 Netty 服务同时监听多个端口。Worker 线程数默认是 CPU 核数的两倍,Boss 线程主要监听 SocketChannel 的 OP_ACCEPT 事件和客户端的连接(主线程)。

当 Boss 线程监听到有 SocketChannel 连接接入时,会把 SocketChannel 包装成 NioSocketChannel,并注册到 Worker 线程的 Selector 中,同时监听其 OP_WRITE 和 OP_READ 事件。当 Worker 线程监听到某个 SocketChannel 有就绪的读 I/O 事件时,会进行以下操作。

(1)向内存池中分配内存,读取 I/O 数据流。

(2)将读取后的 ByteBuf 传递给解码器 Handler 进行解码,若能解码出完整的请求数据包,就会把请求数据包交给业务逻辑处理 Handler。

(3)经过业务逻辑处理 Handler 后,在返回响应结果前,交给编码器进行数据加工。

(4)最终写到缓存区,并由 I/O Worker 线程将缓存区的数据输出到网络中并传输给客户端。

Netty 主从线程模型如图 2-3 所示,图中有个任务队列,这个任务队列主要是用来处理一些定时任务的,如连接的心跳检测。同时,当开启了额外业务线程时,写回响应结果也会被封装成任务,交给 I/O Worker 线程来完成。

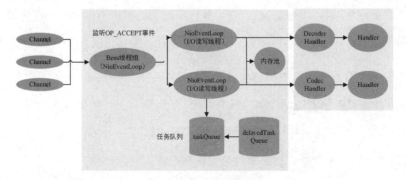

图 2-3 Netty 主从线程模型

2.3 编码和解码

在第 1 章介绍 Netty 客户端与服务端的通信原理时，使用过编码器和解码器，但并未对其底层原理进行详细的介绍。如果使用 Java NIO 来实现 TCP 网络通信，则需要对 TCP 连接中的问题进行全面的考虑，如拆包和粘包导致的半包问题和数据序列化等。对于这些问题，Netty 都做了很好的处理。本节通过 Netty 的编码和解码架构及其源码对上述问题进行详细剖析。下面先看一幅简单的 TCP 通信图，如图 2-4 所示。

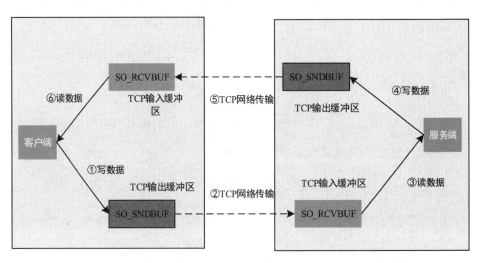

图 2-4　TCP 通信图

在图 2-4 中，客户端给服务端发送消息并收到服务端返回的结果，共经历了以下 6 步。

① TCP 是面向字节流传输的协议，它把客户端提交的请求数据看作一连串的无结构的字节流，并不知道所传送的字节流的含义，也并不关心有多少数据流入 TCP 输出缓冲区中。

② 每次发多少数据到网络中与当前的网络拥塞情况和服务端返回的 TCP 窗口的大小有关，涉及 TCP 的流量控制和阻塞控制，且与 Netty 的反压有关。如果客户端发送到 TCP 输出缓冲区的数据块太多，那么 TCP 会分割成多次将其传送出去；如果太少，则会等待积累足够多的字节后发送出去。很明显，TCP 这种传输机制会产生粘包问题。

③ 当服务端读取 TCP 输入缓冲区中的数据时，需要进行拆包处理，并解决粘包和拆包问题，比较常用的方案有以下 3 种。

- 将换行符号或特殊标识符号加入数据包中，如 HTTP 和 FTP 等。
- 将消息分为 head 和 body，head 中包含 body 长度的字段，一般前面 4 个字节是 body 的长度值，用 int 类型表示，但也有像 Dubbo 协议那种，head 中除 body 长度外，还有版本号、请求类型、请求 id 等。
- 固定数据包的长度，如固定 100 个字节，不足补空格。

步骤④～⑥与步骤①～③类似。TCP 的这些机制与 Netty 的编码和解码有很大的关系。Netty 采用模板设计模式实现了一套编码和解码架构，高度抽象，底层解决 TCP 的粘包和拆包问题，对前面介绍的 3 种方案都做了具体实现。

第 1 种方案，Netty 有解码器 LineBasedFrameDecoder，可以判断字节中是否出现了 "\n" 或 "\r\n"。

第 2 种方案，Netty 有编解码器 LengthFieldPrepender 和 LengthFieldBasedFrameDecoder，可以在消息中加上消息体长度值，这两个编解码器在之前的实战中用到过。

第 3 种方案，Netty 有固定数据包长度的解码器 FixedLengthFrameDecoder。此方案一般用得较少，比较常用的是前两种方案。

Netty 对编码和解码进行了抽象处理。编码器和解码器大部分都有共同的编码和解码父类，即 MessageToMessageEncoder 与 ByteToMessageDecoder。ByteToMessageDecoder 父类在读取 TCP 缓冲区的数据并解码后，将剩余的数据放入了读半包字节容器中，具体解码方案由子类负责。在解码的过程中会遇到读半包，无法解码的数据会保存在读半包字节容器中，等待下次读取数据后继续解码。编码逻辑比较简单，MessageToMessageEncoder 父类定义了整个编码的流程，并实现了对已读内存的释放，具体编码格式由子类负责。

Netty 的编码和解码除了解决 TCP 协议的粘包和拆包问题，还有一些编解码器做了很多额外的事情，如 StringEncode（把字符串转换成字节流）、ProtobufDecoder（对 Protobuf 序列化数据进行解码）；还有各种常用的协议编解码器，如 HTTP2、Websocket 等。本节只是介绍了 Netty 为什么要编/解码、Netty 编/解码的实现思想，以及一些常用编/解码的使用，后续章节会对常用的编码器和解码器进行实战应用，并进行详细的源码剖析。

2.4 序列化

在图 2-4 中，当客户端向服务端发送数据时，如果发送的是一个 Java 对象，由于网络只能传输二进制数据流，所以 Java 对象无法直接在网络中传输，则必须对 Java 对象的内容进行流化，因为只有流化后的对象才能在网络中传输。序列化就是将 Java 对象转换成二进制流数据的过程，而这种转化方式多种多样，本节介绍几种常用的序列化方式。

（1）Java 自带序列化：使用非常简单，但在网络传输时很少使用，这主要是因为其性能太低，序列化后的码流太大。另外，它无法跨语言进行反序列化。

（2）为了解决 Java 自带序列化的缺点，会引入比较流行的序列化方式，如 Protobuf、Kryo、JSON 等。由于 JSON 格式化数据可读性好，而且浏览器对 JSON 数据的支持性非常好，所以一般的 Web 应用都会选择它。另外，市场上有 Fastjson、Jackson 等工具包，使 Java 对象转换成 JSON 也非常方便。但 JSON 序列化后的数据体积较大，不适合网络传输和海量数据存储。Protobuf 和 Kryo 序列化后的体积与 JSON 相比要小很多，本节会对这两种序列化各自的优点和缺点及应用场景进行详细的讲解。

2.4.1 Protobuf 序列化

Protobuf 是 Google 提供的一个具有高效协议数据交换格式的工具库（类似 JSON），但 Protobuf 有更高的转化效率，且时间效率和空间效率都是 JSON 的 3～5 倍，为何采用 Protobuf 序列化后占用的存储空间要比 JSON 占用的存储空间小而反序列化要更快呢？例如，有一个 Java 对象，将其转换为 JSON 格式{ "userName": "zhangsan", "age": 20, "hobby": "swimming" }，很显然，当采用这种方式进行序列化时，会写进去一些无用的信息，如{"userName",…}，此实例对象字段少，就算再怎么浪费也无妨。但当类的属性非常多并包含各种对象组合时，开销会非常大，有时甚至会超过真正需要传送的值。

Protobuf 对这些字段属性进行了额外处理，同类的每个属性名采用 Tag 值来标识，这个 Tag 值在 Protobuf 中采用了 varint 编码，当类的属性个数小于 128 时，每个属性名只需 1B 即可表示，同时属性值的长度也只占用 1B。Protobuf 对值也进行了各种编码，不同类型的数据值采用不同的编码技术，以尽量减小占用的存储空间。可以将 Protobuf 序列化后的数据想象成下面这样的格式：

| tag | length | value | tag | length | value | … | tag | length | value |

Protobuf 序列化除了占用空间小，性能还非常好，主要是它带有属性值长度，无须进行字符串匹配，这个长度值只占用 1B 的存储空间。另外，JSON 都是字符串解析，而 Protobuf 根据不同的数据类型有不同的大小，如 bool 类型只需读取 1B 的数据。

Protobuf 的缺点如下。

（1）从 Protobuf 序列化后的数据中发现，Protobuf 序列化不会把 Java 类序列化进去。当遇到对象的一个属性是泛型且有继承的情况时，Protobuf 序列化无法正确地对其进行反序列化，还原子类信息。

（2）Protobuf 需要编写 .proto 文件，比较麻烦，此时可以使用 Protostuff 来解决。Protostuff 是 Protobuf 的升级版，无须编写.proto 文件，只需在对象属性中加入@Tag 注解即可。

（3）可读性差，只能通过程序反序列化解析查看具体内容。

本小节对 Protobuf 有了一个初步的了解。Protobuf 一般用于公司内部服务信息的交换。目前市场上序列化框架有很多，了解其优点和缺点对技术选型有很大的帮助。对于数据量比较大、对象属性是无泛型且有继承的数据，Protobuf 是个很不错的工具，值得推荐。

2.4.2　Kryo 序列化

本小节介绍另外一个序列化框架——Kryo，它是一个快速高效的 Java 对象图形序列化框架，主要特点是性能高、高效和易用。它的存储与 Protobuf 的存储一样，采用了可变长存储机制。Kryo 的序列化方式比较常用的有以下两种。

（1）默认序列化方式——FieldSerializer。此方式非常高效，只写字段数据，没有任何额外信息，但是它不支持添加、删除或更改字段类型，并不适合业务系统，因为业务数据模型会经常改变。它只适用于序列化与反序列化的相关类一致的情况。

（2）在长期持久化存储方面，可以考虑 Kryo 的另一种序列化方式——TaggedFieldSerializer。TaggedFieldSerializer 与 Protobuf 非常相似，在需要每个字段上加上@Tag 注解，不仅可以修改字段，还支持泛型的继承，但会写入对象中用到的具体类型。Kryo 还支持对类进行注册，为每个类分配一个 id，用一个字节表示一长串的类名。但这种方式很容易出现问题，因此业务系统用 Kryo 基本上不考虑。

在使用 Kryo 时，需要仔细考虑以下几点。

- 当采用 Tag 方式兼容修改数据时要注意 Kryo 的版本。例如，选择 Kryo 4.0.2 版本，除了设置 setSkipUnknownTags(true)，还需要在@Tag 注解里加上 annexed=true，如@Tag(value=1, annexed=true)，当前字段在反序列化时才可以删除；Kryo 4.0.0 版本则无须在@Tag 注解里加上 annexed=true。
- Kryo 的引用机制。若开启了引用，则字段值除了基本类型，其他的都有可能被引用上。例如，字符串的 a 字段值如果是""空字符串，b 字段值和 a 字段值一样，则在反序列化时，a 字段在类中被删除，引用会报错（会报数组越界的异常）。如果系统无循环引用，则无须开启引用。
- 使用无须加注解的兼容模式——CompatibleFieldSerializer 序列化方式。这种方式虽然能支持泛型的继承，但序列化后的数据占用的空间可能会比 JSON 序列化占用的空间还大，因此一般不建议使用。
- Kryo 是非线程安全的，可以采用 ThreadLocal 将其缓存起来。

上述 Kryo 的注意细节大部分是编者在使用过程中曾经遇到的问题，具体实战还需进入 Kryo 官网仔细阅读。

2.5 零拷贝

序列化主要与传输数据格式有关，不管是 Kryo 还是 Protobuf，它们都能对数据内容进行压缩，并能完整地恢复。零拷贝是 Netty 的一个特性，主要发生在操作数据上，无须将数据 Buffer 从一个内存区域拷贝到另一个内存区域，少一次拷贝，CPU 效率就会提升。Netty 的零拷贝主要应用在以下 3 种场景中。

（1）Netty 接收和发送 ByteBuffer 采用的都是堆外直接内存，使用堆外直接内存进行 Socket 的读/写，无须进行字节缓冲区的二次拷贝。如果使用传统的堆内存进行 Socket 的读/写，则 JVM 会将堆内存 Buffer 数据拷贝到堆外直接内存中，然后才写入 Socket 中。与堆外直接内存相比，使用传统的堆内存，在消息的发送过程中多了一次缓冲区的内存拷贝。

（2）在网络传输中，一条消息很可能会被分割成多个数据包进行发送，只有当收到一个完整的数据包后，才能完成解码工作。Netty 通过组合内存的方式把这些内存数据包逻辑组合到一块，而不是对每个数据块进行一次拷贝，这类似于数据库中的视图。CompositeByteBuf 是 Netty 在此零拷贝方案中的组合 Buffer，在第 4 章节会对它进行详细剖析。

（3）传统拷贝文件的方法需要先把文件采用 FileInputStream 文件输入流读取到一个临时的 byte[]数组中，然后通过 FileOutputStream 文件输出流，把临时的 byte[]数据内容写入目的文件中。当拷贝大文件时，频繁的内存拷贝操作会消耗大量的系统资源。Netty 底层运用 Java NIO 的 FileChannel.transfer()方法，该方法依赖操作系统实现零拷贝，可以直接将文件缓冲区的数据发送到目标 Channel 中，避免了传统的通过循环写方式导致的内存数据拷贝问题。

本节内容偏少，且理论性较强。零拷贝机制在做上层应用时几乎不会接触，但在面试时，很有可能被问到，因此一定要深入理解这 3 种重要的零拷贝场景。如果目前难以理解，则可暂时跳过，在仔细看完源码剖析部分内容后，再来仔细分析。

2.6 背压

本节是 Netty 基本理论知识的最后一节，主要介绍如何运用 Netty 实现背压，并在此基础上分析它的具体应用实例，不管是在面试时，还是在实际工作中，都是非常有用的。下面先看一下 TCP 链路背压场景图，如图 2-5 所示。

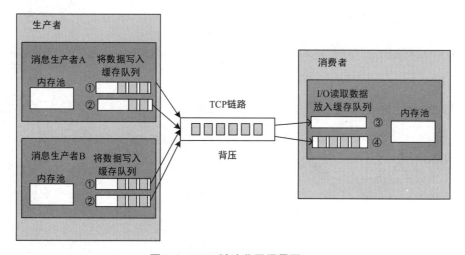

图 2-5　TCP 链路背压场景图

细看图 2-5 可以发现，当消费者的消费速率低于生产者的发送速率时，会造成背压，此时消费者无法从 TCP 缓存区中读取数据，因为它无法再从内存池中获取内存，从而造成 TCP 通道阻塞。生产者无法把数据发送出去，这就使生产者不再向缓存队列中写入数据，从而降低了生产速率。当消费者的消费速率提升且 TCP 通道不再阻塞时，生产者的发送速率又会得到提升，整个链路运行恢复正常。

2.6.1　TCP 窗口

Netty 的背压主要运用 TCP 的流量控制来完成整个链路的背压效果，而在 TCP 的流量控制中有个非常重要的概念——TCP 窗口。TCP 窗口的大小是可变的，因此也叫滑动窗口。TCP 窗口本质上就是描述接收方的 TCP 缓存区能接收多少数据的，发送方可根据这个值来计算最多可以发送数据的长度。接下来看图 2-6，以了解 TCP 窗口的工作过程，图中涉及 TCP 窗口大小和 ACK，这些参数都是 TCP 头部比较重要的概念。

- Sequence Number：包的序号，用来解决网络包乱序问题。
- Acknowledgement Number：简称 ACK（确认序号），用来解决丢包问题。
- Window：TCP 窗口，也叫滑动窗口，用来解决流控。
- TCP Flag：包的类型，主要用来操控 TCP 的 11 种状态机。

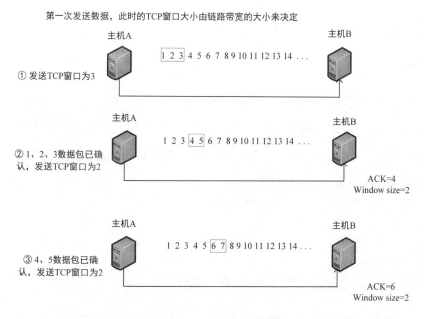

图 2-6　TCP 窗口的工作过程

在图 2-6 中，主机 A 与主机 B 通信，在第一次发送数据时，主机 A 发送多少数据到网络中由链路带宽的大小来决定。同时，主机 B 在收到数据段后，会把下一个要接收的数据序号返回给主机 A，即图中的 ACK。

（1）例如，主机 A 发送 3 个数据段长度，然后等待主机 B 的确认，当主机 B 收到 1～3 数据段时，会给主机 A 返回 ACK=4，以及当前的 TCP 窗口大小（这个窗口大小为接收端可用窗口），此时 TCP 窗口大小为 2。

（2）主机 A 在收到这些信息后，根据 TCP 窗口大小和主机 B 期待收到下一个字节的确认序号滑动其窗口,此时 TCP 窗口包含 4～5 数据段。同理继续发送 4～5 和 6～7 数据段。

(3)当 TCP 窗口大小为 0 时,主机 A 发送端将停止发送数据,直到 TCP 窗口大小恢复为非零值。

一些分布式计算引擎主要是利用了 TCP 窗口的特性,因此,无须加太多额外处理流程就能完成整套架构的背压,如 Flink。

2.6.2 Flink 实时计算引擎的背压原理

Flink 采用 Netty 发送数据时的高低水位来控制整个链路的背压,是非常好的 Netty 背压实现实例。读者可能会思考,为何像 Dubbo 这种 RPC 框架不采用 Netty 发送数据时的高低水位来控制整个链路的背压呢?这是因为 RPC 框架一般只需做请求 TPS 流量控制即可,而 Flink 有生产者和消费者,若消费者处理能力非常弱,则生产者需要得到感应,而且此时需要降低生产速率,以缓解消费者的压力。因此索引 Flink 无法通过限流来控制整个链路,需要用背压机制来解决。

在 Flink 中运行的作业主要分以下三种。

第一种:用于接收数据,如接收 Kafka 的数据源。

第二种:Task 对这些数据进行处理,如 map、reduce、filter、join 等操作,在这些操作过程中,可能会涉及 HBase 等数据库的读/写操作。

第三种:Task 主要对第二种 Task 计算的结果数据进行输出,如输出到 Kafka、HBase 中等。

在高峰期,接收数据的速率远高于处理数据的速率。例如,当程序更新版本时,需要停止服务,数据在 Kafka 上会积压一段时间。当服务启动时,流入的数据会快速堆积。此时,如果 Flink 没有背压,则可能会导致处理数据的 TaskManager 内存耗尽,甚至整个系统直接崩溃。

Flink 究竟是如何处理背压的呢?它是如何运用 Netty 发送数据时的高低水位及 TCP 的流

量控制来实现整条链路的背压的呢？下面来看 Flink Task 之间通信时的内存分配与管理，如图 2-7 所示。

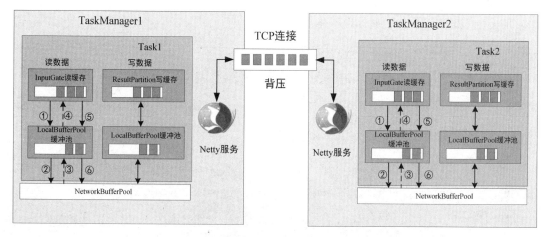

图 2-7　Flink Task 之间通信时的内存分配与管理

考虑到很多 Java 程序员并未接触过 Flink、JStorm 等大数据框架，因此不对其内部架构与原理进行详细的讲解。可以把图 2-7 中的 TaskManager 看作一个 JVM 服务，把 Task 看作一条线程，一个 JVM 服务中运行多条 Task 线程，Task 之间通过 Netty 长连接进行数据交互。当然，一个 TaskManager 中只有一个 Netty，当 Netty 接收到数据后，它会把数据拷贝到 Task 中，拷贝数据需要内存。图 2-7 中的 Task2 作为消费者，其内存的申请及背压处理步骤如下。

（1）先到对应 Channel 的 LocalBufferPool 缓冲池中进行申请，若缓冲池中没有可用的内存，且已申请的数量还未达到缓冲池的上限，则向 NetworkBufferPool 申请内存块，即图中的①和②。

（2）申请成功后，将其交给 Channel 填充数据，即图中的③和④。

（3）若 LocalBufferPool 缓冲池申请的数量已经达到上限或 NetworkBufferPool 中的内存已经被用尽，那么当前 Task 的 Netty Channel 暂停读取数据。此时数据积压在 TCP 缓冲区中，导致其 TCP 窗口的大小变成零，其上游发送端会立刻暂停发送，整个链路进入背压状态。

（4）在 Task1 中，当写数据到达 ResultPartition 写缓存中时，也会向 LocalBufferPool 缓冲

池请求内存块，如果没有可用内存块，则线程状态变成 TIME_AWAIT，并且每隔一段时间就会去申请一次，整条线程也会阻塞在请求内存块的地方，达到暂停写入的目的。

那么，问题来了，背压形成后，整条链路什么时候才能恢复正常？想要恢复正常，只需 Task 线程在写数据时能正常申请到内存即可。当 Task2 输出端的处理能力增强后，会调用内存回收方法，将内存块还给 LocalBufferPool 缓冲池。如果 LocalBufferPool 缓冲池中当前申请的数量达到了上限，那么它会将该内存块回收给 NetworkBufferPool，即图 2-7 中的⑤和⑥，TCP 窗口也会慢慢恢复正常。

通过对图 2-7 进行的详细分析可知，Flink 的背压好像与 Netty 发送数据时的水位控制没太大关系。但是，为了保证服务的稳定，Flink 在数据生产端使用 Netty 发送数据时的高低水位机制来控制整个链路的背压，不向缓存中写太多数据。如果 Netty 输出缓冲区的字节数超过了高水位值，则 Channel.isWritable()为 false，会触发 Handler 的 channelWritabilityChanged()方法。Flink 在发送数据时，若发现 Channel.isWritable()为 false，则不会从发送队列中 poll 出需要发送的数据，从而形成背压。当 Netty 输出缓冲区的字节数降到低水位值以下时，Channel.isWritable()返回 true，同时 channelWritabilityChanged()方法被触发，Flink 在 channelWritabilityChanged 事件中调用发送方法，这样可以继续发送数据。

Netty 的水位值设置如下：

```
bootstrap.childOption(ChannelOption.WRITE_BUFFER_LOW_WATER_MARK,
                config.getMemorySegmentSize() + 1);
bootstrap.childOption(ChannelOption.WRITE_BUFFER_HIGH_WATER_MARK,
                2 * config.getMemorySegmentSize());
```

第 5 章会对 Netty 的高低水位进行详细的讲解。本小节只是讲了一些 Flink 的内存管理及其主要背压原理和思想，内存管理对后续理解 Netty 的内存管理有一定的帮助。例如，在图 2-7 中，既然有了 NetworkBufferPool，为何还要用到对应 Channel 的 LocalBufferPool 缓冲池？这与 Netty 有了 PoolArena，还需要用到 PoolThreadLocalCache 类似。目的是为每个 Channel 预先分配内存，减少内存在分配过程中的多线程竞争，提高性能。

2.7 小结

本章大部分内容都是 TCP 和 NIO 的理论知识，主要涉及数据序列化、编/解码器，以及数据的读/写和传输。本章内容大部分在平时工作中很少接触到，但对了解 Netty 底层原理有很大的帮助。想要写出高性能的代码，就必须对其底层原理有很深入的了解。

第 3 章

分布式 RPC

本章运用 Netty 实现了一套分布式 RPC 服务。结合第 1 章对客户端和 Netty 服务端的长连接通信，想要实现完整的分布式 RPC，需要完成以下两项工作。

（1）在编写业务代码时，只需像运用 SpringMVC 或 Dubbo 一样，引入对应的 jar 包即可，无须关注太多的 Netty 底层实现。

（2）服务端可动态扩展，不用指定具体的 IP 地址和端口进行连接通信。

当服务端与客户端通信时，需要制定上层协议，运用 Java 反射机制，把协议内容与代码进行映射，让业务代码与 Netty 的 Handler 逻辑处理解耦。同时引入分布式协调器 Zookeeper，实现服务的注册与发现，动态扩展服务。

3.1　Netty 整合 Spring

Web 应用程序开发一般都会引入 Spring 框架，在 RPC 框架实现前需要先整合 Spring 容器。Spring 的整合需要考虑如何调用 Netty 服务及启动 Netty 服务监听端口。

Netty 服务启动后会阻塞线程，因此可以通过新建线程去启动它。由于 Netty 服务启动后会使用容器中的 Bean，所以只有在 Spring 容器把所有的 Bean 初始化完成后才能去启动。Spring 容器中有一种监听器，可以在监听到 Bean 初始化完成后得到触发，这种监听器需要实现 ApplicationListener<ContextRefreshedEvent>接口，在其 onApplicationEvent()方法里启动 Netty 服务：先在 pom.xml 文件中引入 Spring 5.1.9.RELEASE 版本的依赖。具体实现代码如下：

```xml
<dependency>
    <groupId>org.springframework</groupId>
    <artifactId>spring-core</artifactId>
    <version>5.1.9.RELEASE</version>
</dependency>
<dependency>
    <groupId>org.springframework</groupId>
    <artifactId>spring-beans</artifactId>
    <version>5.1.9.RELEASE</version>
</dependency>
<dependency>
    <groupId>org.springframework</groupId>
    <artifactId>spring-context</artifactId>
    <version>5.1.9.RELEASE</version>
</dependency>
```

编写 Spring 容器启动类 ApplicationMain。采用扫描注解方式启动，并为 Runtime 添加关闭钩子函数，实现优雅停机。具体代码如下：

```java
package com.itjoin.pro_netty.spring;
import org.springframework.context.annotation.AnnotationConfigApplicationContext;
public class ApplicationMain {
```

```java
private static volatile boolean running = true;
public static void main(String[] args) {
    try {
        //注意,扫描包名时要小心,只扫描服务端用到的包路径
        AnnotationConfigApplicationContext context = new
        AnnotationConfigApplicationContext("com.itjoin");
        //在 JVM 中增加一个关闭的钩子,当 JVM 关闭时
        //会执行系统中已经设置的所有通过方法 addShutdownHook()添加的钩子
        //只有当系统执行完这些钩子后,JVM 才会关闭
        Runtime.getRuntime().addShutdownHook(new Thread() {
            public void run() {
                try {
                    context.stop();
                } catch (Throwable t) {
                }
                synchronized (ApplicationMain.class) {
                    running = false;
                    ApplicationMain.class.notify();
                }
            }
        });
        context.start();
    } catch (Exception e) {
        e.printStackTrace();
        System.exit(1);
    }
    System.out.println("服务器已启动====");
    synchronized (ApplicationMain.class) {
        while (running) {
            try {
                ApplicationMain.class.wait();
            } catch (Throwable e) {
            }
        }
    }
}
```

            }
}

在 Spring 监听器类 NettyApplicationListener 的 onApplicationEvent()方法中新建线程，启动 Netty 服务，然后把 Netty 服务中 main()方法的代码放到 start()方法中，即完成了 Spring 的整合。具体代码如下：

```java
package com.itjoin.pro_netty.spring;
import com.itjoin.pro_netty.server.Netty服务;
import org.springframework.context.ApplicationListener;
import org.springframework.context.event.ContextRefreshedEvent;
import org.springframework.stereotype.Component;
@Component
public class NettyApplicationListener implements
 ApplicationListener<ContextRefreshedEvent> {
    @Override
    public void onApplicationEvent(ContextRefreshedEvent contextRefreshedEvent) {
            //开启额外线程，启动 Netty 服务
             new Thread(){
            @Override
            public void run() {
                Netty服务.start();
            }
        }.start();
    }
}
```

3.2 采用 Netty 实现一套 RPC 框架

虽然完成了 Spring 的整合，但只是采用 Spring 容器启动了 Netty 服务。本节采用 Netty 实现一套 RPC 框架，微服务统一使用这套框架来实现 RPC 通信。在改造代码之前，先看一幅 RPC 内部消息处理图，如图 3-1 所示。

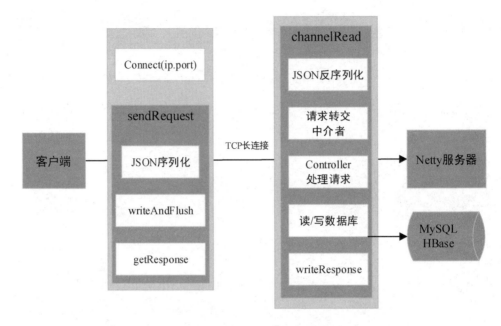

图 3-1 RPC 内部消息处理图

在图 3-1 中，除了请求转交中介者和 Controller 处理请求并执行业务代码读/写数据库，大部分功能在第 1 章都已实现。在平时的工作中，当运用 SpringMVC 框架编写业务代码时，只需新建 Controller 类和对应的接口方法即可，无须知道 SpringMVC 的底层实现逻辑。接口实现完成后，通过 HTTP 请求指定 URL 来实现远程调用。运用 Netty 实现的 RPC 与运用 SpringMVC 实现的 RPC 类似。但对于通信协议，本书只支持 TCP。

SpringMVC 底层主要通过解析 URL 获取 Controller 对象和对应的接口方法，然后运用 Java 的反射运行对应的接口方法。获取的方式依赖@Controller 和@RequestMapping 注解，由 URL 解析出接口方法上的注解@RequestMapping 的值，再根据这个值映射对应的接口方法。这种映射方式需要把注解@RequestMapping 的值放入 Map 容器中缓存起来，Map 中的 key 为注解 @RequestMapping 的值、value 为对应的接口方法的 Method 对象。当读取 URL 时，就相当于有了 key，此时就可以从容器中获取接口方法的 Method 对象了。

本书 RPC 框架的实现借用的也是上述这种方式，只是具体路径需要客户端和服务端制定上层协议。客户端每次在发送请求时都需要把请求路径传送给服务端,服务端获取路径后,在本地缓存 Map 中得到对应的调用方法。本地缓存 Map 在构建之前需要先编写注解类@Remote(与 SpringMVC 中的@RequestMapping 注解类似),这个注解作用于接口方法,通过扫描这个注解类,可以获取所有的接口方法。具体代码如下:

```
package com.itjoin.pro_netty.annotation;
import java.lang.annotation.ElementType;
import java.lang.annotation.Retention;
import java.lang.annotation.RetentionPolicy;
import java.lang.annotation.Target;
//此注解只能放在方法上
@Target({ElementType.METHOD})
@Retention(RetentionPolicy.RUNTIME)
public @interface Remote {
    String value();
}
```

构建本地缓存 Map 容器 Mediator.methodBeans,用于缓存所有接口的对象和方法。把容器放在中介者 Mediator 类中,这个类把 Netty 代码与业务代码解耦,后面还包含协议的解析,接口方法的调用。具体代码如下:

```
import java.lang.reflect.Method;
import java.util.HashMap;
import java.util.Map;
public class Mediator {
    public static Map<String, MethodBean> methodBeans;
    static{
        methodBeans = new HashMap<>();
    }
    public static class MethodBean {
        private Object bean;
        private Method method;
        public Object getBean() {
            return bean;
        }
        public void setBean(Object bean) {
```

```
        this.bean = bean;
    }
    public Method getMethod() {
        return method;
    }
    public void setMethod(Method method) {
        this.method = method;
    }
}
```

当 Spring 容器启动并完成 Bean 的初始化后,可以运用上下文刷新事件 ContextRefreshedEvent,在事件中循环遍历容器中的 Bean,获取带有 Controller 的注解对象及其 @Remote 注解方法。并把它们放入缓存容器 Mediator.methodBeans 中。由于 Netty 服务的启动也是在 ContextRefreshedEvent 事件中完成的,所以两个动作的执行有先后顺序,为了保证在 Netty 服务启动前所有接口方法都已放入缓存容器中,Spring 容器提供了 Ordered 接口,用来处理相同接口实现类的优先级问题。具体实现类 InitLoadRemoteMethod 的代码如下:

```
package com.itjoin.pro_netty.spring;
import com.itjoin.pro_netty.annotation.Remote;
import com.itjoin.pro_netty.core.Mediator;
import org.springframework.context.ApplicationListener;
import org.springframework.context.event.ContextRefreshedEvent;
import org.springframework.core.Ordered;
import org.springframework.stereotype.Component;
import org.springframework.stereotype.Controller;
import java.lang.reflect.Method;
import java.util.Map;
/**
 * Spring 容器初始化后,把带有@Remote 的方法与其对象加载到缓存中
 * Ordered 接口主要通过 getOrder()的返回值来决定监听器运行的顺序
 * getOrder()的返回值越小,运行顺序越靠前,此处为-1
 * 这是由于初始化 Remote()方法需要在 Netty 服务启动之前
 */
@Component
public class InitLoadRemoteMethod implements
```

```java
ApplicationListener<ContextRefreshedEvent> , Ordered {
    @Override
    public void onApplicationEvent(ContextRefreshedEvent contextRefreshedEvent) {
        //从 Spring 容器中获取标有 Controller 注解的对象
        Map<String, Object> controllerBeans
            = contextRefreshedEvent.getApplicationContext()
                .getBeansWithAnnotation(Controller.class);
        //遍历所有的 Controller
        for(String key : controllerBeans.keySet()){
            Object bean = controllerBeans.get(key);
            //通过反射获取 Controller 的所有方法
            Method[] methods = bean.getClass().getDeclaredMethods();
            for (Method method : methods){
                //判断方法上是否有@Remote 注解
                if(method.isAnnotationPresent(Remote.class)){
                    //获取@Remote 注解上的 value 值
                    Remote remote = method.getAnnotation(Remote.class);
                    String methodVal = remote.value();
                    //把方法和 Bean 放入包装类 MethodBean 中
                    Mediator.MethodBean methodBean = new
                    Mediator.MethodBean();
                    methodBean.setBean(bean);
                    methodBean.setMethod(method);
                    //最终把@Remote 注解里的值作为 key,将方法和 bean 包装好并作为 value 放入本地缓存中
                    Mediator.methodBeans.put(methodVal,methodBean);
                }
            }
        }
    }
    @Override
    public int getOrder() {
        return -1;//值越小优先级越高
    }
}
```

缓存容器 Mediator.methodBeans 初始化后，中介者 Mediator 需要根据 RequestFuture 请求从缓存容器中获取接口方法。RequestFuture 类加上路径 String path 属性，服务端 Mediator 根

据 path 的值从缓存中获取调用对象和方法，运用 Java 反射运行业务逻辑处理方法并获取执行结果。方法参数类型对 List 泛型集合需要用 JSONArray 反序列化。Mediator 类的最终代码如下：

```java
package com.itjoin.pro_netty.core;
import com.alibaba.fastjson.JSONArray;
import com.alibaba.fastjson.JSONObject;
import com.itjoin.pro_netty.asyn.RequestFuture;
import com.itjoin.pro_netty.asyn.Response;
import java.lang.reflect.Method;
import java.util.HashMap;
import java.util.List;
import java.util.Map;
/**
 * 此类主要对Netty网络通信与业务处理逻辑类起到沟通的关联作用
 */
public class Mediator {
    public static Map<String, MethodBean> methodBeans;
    static{
        methodBeans = new HashMap<>();
    }
    /**
     * 请求分发处理
     * @param requestFuture
     * @return
     */
    public static Response process(RequestFuture requestFuture){
        //构建响应对象
        Response response = new Response();
        try {
            String path = requestFuture.getPath();
            //根据请求路径从缓存中获取请求路径对应的bean和method
            MethodBean methodBean = methodBeans.get(path);
            if(methodBean!=null){
                Object bean = methodBean.getBean();
                Method method = methodBean.getMethod();
                //获取请求内容
```

```
                Object body = requestFuture.getRequest();
                //获取方法的请求参数类型,此处只支持一个参数
                //想支持多个参数需要进行相应的修改
                Class[] paramTypes = method.getParameterTypes();
                Class paramType = paramTypes[0];
                Object param = null;
                //如果参数是 List 类型
                if(paramType.isAssignableFrom(List.class)){
                //采用 JSONArray 反序列化
                param = JSONArray.
                        parseArray(JSONArray.toJSONString(body),paramType);
                //如果参数是 String 类型
                }else if(paramType.getName().
                equals(String.class.getName())){
                    param = body;
                }else{
                    //采用 JSONObject 反序列化
                param
                        = JSONObject.parseObject(JSONObject.toJSONString(body),paramType);
                }
                //采用 Java 反射运行业务逻辑处理方法并获取返回结果
                Object result = method.invoke(bean,param);
                response.setResult(result);
            }
        } catch (Exception e) {
            e.printStackTrace();
        }
        response.setId(requestFuture.getId());
        return response;
    }
    public static class MethodBean {
        private Object bean;
        private Method method;
        public Object getBean() {
            return bean;
        }
        public void setBean(Object bean) {
```

```
        this.bean = bean;
    }
    public Method getMethod() {
        return method;
    }
    public void setMethod(Method method) {
        this.method = method;
    }
}
```

Mediator 类作为中介者,衔接 Netty 服务的 Handler 类和业务逻辑处理类。如果需要对服务器的 Handler 进行一些改动,就引入 Mediator,并把请求 RequestFuture 交给它去处理。

```
package com.itjoin.pro_netty.server;
import java.nio.charset.Charset;
import com.alibaba.fastjson.JSONObject;
import com.itjoin.pro_netty.asyn.RequestFuture;
import com.itjoin.pro_netty.asyn.Response;
import com.itjoin.pro_netty.core.Mediator;
import io.netty.buffer.ByteBuf;
import io.netty.channel.ChannelHandlerContext;
import io.netty.channel.ChannelInboundHandlerAdapter;
//@Sharable 注解表示此 Handler 对所有 Channel 共享,无状态,注意多线程并发
@ChannelHandler.Sharable
public class ServerHandler extends ChannelInboundHandlerAdapter {
    /**
     * 读取客户端发送的数据
     */
    @Override
    public void channelRead(ChannelHandlerContext ctx, Object msg) {
//if(msg instanceof ByteBuf) {
//ByteBuf 的 toString()方法把二进制数据转换成字符串,默认编码 UTF-8
//System.out.println(((ByteBuf)msg).toString(Charset.defaultCharset()));
//}
//ctx.channel().writeAndFlush("msg has recived!");
        //获取客户端发送的请求,并将其转换成 RequestFuture 对象
        //由于经过了 StringDecoder 解码器,所以 msg 为 String 类型
```

```
        RequestFuture request
            = JSONObject.parseObject(msg.toString(),RequestFuture.class);
//获取请求id
//long id = request.getId();
//System.out.println("请求信息为==="+msg.toString());
//构建响应结果
//Response response = new Response();
//response.setId(id);
//response.setResult("服务器响应ok"+id);
        Response response = Mediator.process(request);
        //把响应结果返回客户端
        ctx.channel().writeAndFlush(JSONObject.toJSONString(response));
    }
    @Override
    public void channelWritabilityChanged(ChannelHandlerContext ctx) throws
      Exception {
    }
}
```

服务端的核心代码基本上改造完了,最后编写一个接口实例 UserController 类,这个类中有根据 userId 参数获取用户信息的方法。具体代码如下:

```
package com.itjoin.pro_netty.controller;
import com.itjoin.pro_netty.annotation.Remote;
import org.springframework.stereotype.Controller;
@Controller
public class UserController {
    //此方法用于远程调用,需要加上@Remote注解
    @Remote("getUserNameById")
    public Object getUserNameById(String userId){
        //此处并未访问数据库,只进行简单的输出
        System.out.println("客户端请求的用户id为======"+userId);
        return "响应结果===用户张三"+userId;
    }
}
```

运行服务端启动类 ApplicationMain,同时修改 NettyClient 的 main()方法,并在 NettyClient

类的 sendRequest()方法中加上 path 参数，把 path 参数 getUserNameById 赋给 RequestFuture 的 path 属性。具体代码如下：

```
public static void main(String[] args) throws ExecutionException {
NettyClient client = new NettyClient();
for(int i=0;i<100;i++) {
  Object result = client.sendRequest("id"+i,"getUserNameById");
  System.out.println(result);
 }
}
```

最终的 RPC Demo 测试输出图如图 3-2 所示。

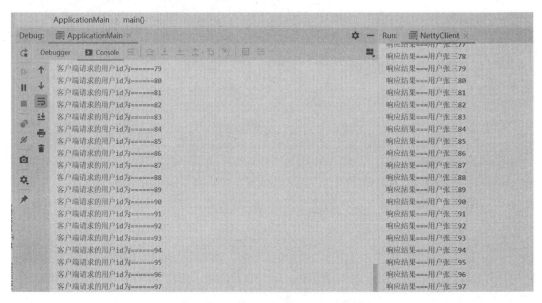

图 3-2　RPC Demo 测试输出图

3.3　分布式 RPC 的构建

RPC 框架整合了 Spring，并实现了 Netty 核心通信代码与业务逻辑代码的解耦，但这个框架需要客户端指定服务端具体的 IP 和端口才能调用，服务无法动态扩展。当然，可以使用 Nginx 或 LVS 等反向代理服务实现横向扩展，但修改反向代理服务器配置并重启无法动态扩展。本节引入一个分布式应用程序协调服务——Zookeeper，以实现服务的注册与发现，完成一套分布式动态扩展 RPC 服务。

3.3.1 服务注册与发现

服务注册与发现是分布式 RPC 必须具备的功能，目前市场上比较常用的服务注册与发现的工具有 Zookeeper、Consul、ETCD、Eureka。

Consul 在这几款产品中功能比较全面，支持跨数据中心的同步、多语言接入、ACL（Access Control Lists，访问控制列表），并提供自身集群的监控。在 CAP 原则的取舍上，Consul 选择了一致性和可用性；Zookeeper 选择了一致性和分区容错性，牺牲了可用性。从理论上来讲，在服务发现的应用场景下，Consul 更合适。

本书选择 Zookeeper 来进行分布式协调，这主要是由于很多大数据组件基本上都会用 Zookeeper，如 Hadoop、HBase、JStorm、Kafka 等。Zookeeper 主要通过心跳来维护活动会话的临时节点，从而维护集群中的服务器状态。分布式 RPC 也是使用 Zookeeper 的临时节点来维护 Netty 服务状态的，RPC 客户端运用 Zookeeper 的 Watch 机制监听 Netty 服务在 Zookeeper 上注册的目录，从而及时感知服务列表的改变。分布式 RPC 架构图如图 3-3 所示。

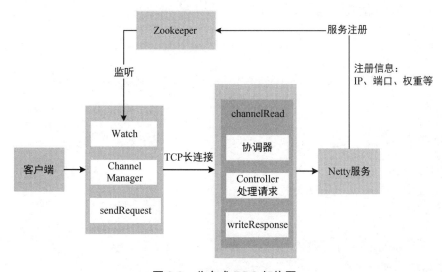

图 3-3　分布式 RPC 架构图

图 3-3 是在普通 RPC 架构的基础上进行了些许改造，在服务端增加了服务注册功能，同时客户端新增了 Watch 机制，用来监听 Netty 服务注册在 Zookeeper 上的目录。客户端与 Netty 服务连接的链路缓存在 ChannelManager 中，一旦发现有 Netty 服务宕机或新增的情况，缓存在 ChannelManager 中的链路就会发生相应的改变。客户端每次在发送请求之前都需要从 ChannelManager 中轮询获取一个连接。

分布式服务发现的负载均衡算法采用的是轮询加权重，每个服务的权重信息都放在配置文件中，当 Netty 服务启动并向 Zookeeper 注册时，需要加上其权重信息。由于 Netty 服务与 Zookeeper 的会话也会出现被断开的情况，所以也需要在服务端加入监听机制。具体实现步骤如下。

（1）修改 Netty 服务，加上服务注册与服务监听 ServerWatcher 类，只要发现其服务本身与 Zookeeper 的临时会话丢失，就需要重新注册。

（2）客户端新增与服务器列表连接的链路管理类 ChannelManager，拥有链路缓存链表，以及对此链表提供增、删、改、查的方法，且必须为原子性操作。

（3）客户端新增监听器 ServerChangeWatcher，监听 Netty 服务注册在 Zookeeper 上的目录，且主要监听其子目录的变化。同时调用 ChannelManager 修改其链路缓存链表。

（4）调整客户端，不再通过服务端 IP 和端口直连，改成从 ChannelManager 中获取连接。

接下来完成这部分代码的开发，先下载 Zookeeper V3.4.6 版本，并进入其 bin 目录，运行 zkServer.cmd，启动 Zookeeper，同时将 Zookeeper 及其 curator 客户端依赖引入 pom.xml 中。

```xml
<dependency>
    <groupId>org.apache.zookeeper</groupId>
    <artifactId>zookeeper</artifactId>
    <version>3.4.12</version>
</dependency>
<dependency>
    <groupId>org.apache.curator</groupId>
    <artifactId>curator-framework</artifactId>
```

```xml
    <version>2.7.1</version>
</dependency>
<dependency>
    <groupId>org.apache.curator</groupId>
    <artifactId>curator-recipes</artifactId>
    <version>2.7.1</version>
</dependency>
```

构建 Zookeeper 连接的工厂类 ZookeeperFactory，它拥有创建 Zookeeper 连接和重连的方法，其中 main()方法只是调试使用。具体代码如下：

```java
package com.itjoin.pro_netty.zookeeper;
import org.apache.curator.RetryPolicy;
import org.apache.curator.framework.CuratorFramework;
import org.apache.curator.framework.CuratorFrameworkFactory;
import org.apache.curator.retry.ExponentialBackoffRetry;
import org.apache.zookeeper.CreateMode;
public class ZookeeperFactory {
    public static CuratorFramework client;
    public static CuratorFramework create(){
        if(client==null){
            RetryPolicy retryPolicy = new ExponentialBackoffRetry(1000, 3);
            //此处连接 Zookeeper 地址，生产环境需要写在配置文件中
            client = CuratorFrameworkFactory.newClient("localhost:2181",
                1000,5000, retryPolicy);
            client.start();
        }
        return client;
    }
    //重新创建连接，主要是为了防止会话丢失时需要重新创建
    public static CuratorFramework recreate(){
        client=null;
        create();
        return client;
    }
    //注意调试，先启动 Zookeeper 服务器
    public static void main(String[] args) throws Exception {
        CuratorFramework client = create();
```

```
        client.create().forPath("/netty");
    }
}
```

Netty 服务类需要连接 Zookeeper 并注册临时节点，还需要监听类 ServerWatcher 监听服务自身与 Zookeeper 的连接。当临时节点丢失时，需要重新创建连接。具体代码如下：

```
public class Netty服务 {
  public static final String SERVER_PATH = "/netty";
  public static void main(String[] args) throws Exception {
    start();
  }
  public static void start() {
    /**
     * 新建两个线程组，即 Boss 线程组与 Worker 线程组
     * 其中，Boss 线程组启动一条线程，用来监听 OP_ACCEPT 事件
     * Worker 线程组默认启动 CPU 核数*2 的线程
     * 用来监听客户端连接的 OP_READ 和 OP_WRITE 事件，并处理 I/O 事件
     */
    EventLoopGroup bossGroup = new NioEventLoopGroup(1);
    EventLoopGroup workerGroup = new NioEventLoopGroup();
    try {
      // ServerBootstrap 为 Netty 服务启动辅助类
      ServerBootstrap serverBootstrap = new ServerBootstrap();
      serverBootstrap.group(bossGroup, workerGroup);
      // 设置 TCP Socket 通道为 NioServerSocketChannel
      // 若是 UDP 通信，则设置为 DatagramChannel
      serverBootstrap.channel(NioServerSocketChannel.class);
      // 设置一些 TCP 参数
      serverBootstrap.option(ChannelOption.SO_BACKLOG, 128)
          /**
           * 当有客户端链路注册读/写事件时，初始化 Handler，并把 Handler 加入管道中
           */
          .childHandler(new ChannelInitializer<SocketChannel>() {
            @Override
            protected void initChannel(SocketChannel ch)
                throws Exception {
              ch.pipeline().addLast(new
```

```java
            LengthFieldBasedFrameDecoder(Integer.MAX_VALUE,
                0, 4, 0, 4));
            // 把接收到的 ByteBuf 数据包转换成 String
            ch.pipeline().addLast(new StringDecoder());
            /**
             * 向 Worker 线程的管道双向链表中添加处理类 ServerHandler
             * 整个处理流向如下：HeadContext-channelRead 读数据-->
             * ServerHandler-channelRead 读取数据做业务逻辑判断
             * 最后将结果传送给客户端
             * -->TailContext-write->HeadContext-write
             */
            ch.pipeline().addLast(new ServerHandler());
            ch.pipeline().addLast(new
             LengthFieldPrepender(4, false));
            // 把字符串消息转换成 ByteBuf
            ch.pipeline().addLast(new
             StringEncoder(Charset.forName("utf-8")));
            // 注意解码器和编码器的顺序
            // 执行顺序正好相反，解码器执行顺序从上往下
            // 编码器执行顺序从下往上
        }
    });
//端口需要放入配置文件中
int port = 8080;
// 同步绑定端口
ChannelFuture future = serverBootstrap.bind(port).sync();
//连接 Zookeeper
CuratorFramework client = ZookeeperFactory.create();
//获取当前服务器的 IP
InetAddress netAddress = InetAddress.getLocalHost();
//当前服务的权重需要放入配置文件中
int weight=1;
//先判断 SERVER_PATH 路径是否存在，若不存在则需要创建
Stat stat =client.checkExists().forPath(SERVER_PATH);
if(stat==null) {
    client.create().creatingParentsIfNeeded()
        .withMode(CreateMode.PERSISTENT).forPath(
```

```java
                    SERVER_PATH,"0".getBytes());
            }
            //在SERVER_PATH目录下构建临时节点，如10.118.15.15#8080#1#0000000000
            client.create().withMode(CreateMode.EPHEMERAL_SEQUENTIAL).
            forPath(SERVER_PATH+"/"+netAddress.getHostAddress()+"#"+
            port+"#"+weight+"#");
            //需要在服务端加上Zookeeper的监控，以防Session中断导致临时节点丢失
            ServerWatcher.serverKey=netAddress.getHostAddress()+port
                                                +Constants.weight;
            client.getChildren().usingWatcher(ServerWatcher.getInstance())
                                            .forPath(SERVER_PATH);
            // 阻塞主线程，直到Socket通道关闭
            future.channel().closeFuture().sync();
        } catch (Exception e) {
            e.printStackTrace();
        } finally {
            // 最终关闭线程组
            workerGroup.shutdownGracefully();
            bossGroup.shutdownGracefully();
        }
    }
}

package com.itjoin.pro_netty.zookeeper;
import com.itjoin.pro_netty.constant.Constants;
import com.itjoin.pro_netty.server.Netty服务;
import org.apache.curator.framework.CuratorFramework;
import org.apache.curator.framework.api.CuratorWatcher;
import org.apache.zookeeper.CreateMode;
import org.apache.zookeeper.WatchedEvent;
import org.apache.zookeeper.Watcher;
import org.apache.zookeeper.data.Stat;
import java.net.InetAddress;
/**
 * 服务端与Zookeeper连接的监听类
 * 以防Session中断导致服务端注册到Zookeeper的临时节点丢失
 */
```

```java
public class ServerWatcher implements CuratorWatcher {
    public static String serverKey = "";
    public static ServerWatcher serverWatcher = null;

    public static ServerWatcher getInstance() {
        if (serverWatcher == null) {
            serverWatcher = new ServerWatcher();
        }
        return serverWatcher;
    }

    /**
     * 监听 Zookeeper 路径的变化
     * 只要有服务器 Session 断了就会触发
     * 当本机与 Zookeeper 的 Session 中断时，需要重新创建临时节点
     * @param event
     * @throws Exception
     */
    @Override
    public void process(WatchedEvent event) throws Exception {
        System.out.println("========服务器监听 Zookeeper=event==="
                +event.getState()+"==="
                +new SimpleDateFormat("yyyy-MM-dd HH:mm:ss").format(new Date()));
        //当会话丢失时
        if (event.getState().equals(Watcher.Event.KeeperState.Disconnected)
                ||event.getState().equals(Watcher.Event.KeeperState.Expired)) {
            try {
                try {
                    //先尝试关闭旧的连接
                    ZookeeperFactory.create().close();
                } catch (Exception e) {
                    e.printStackTrace();
                }
                CuratorFramework client = ZookeeperFactory.recreate();
                client.getChildren().usingWatcher(this).forPath(Netty服务.SERVER_PATH);
                //获取当前服务器的 IP
                InetAddress netAddress = InetAddress.getLocalHost();
```

```
            Stat stat = client.checkExists().forPath(Netty 服务.SERVER_PATH);
            if (stat == null) {
                client.create().creatingParentsIfNeeded()
                        .withMode(CreateMode.PERSISTENT).forPath
                        (Netty 服务.SERVER_PATH,
                        "0".getBytes());
            }
            //在 SERVER_PATH 目录下构建临时节点, 如 10.118.15.15#8080#1#0000000000
            client.create().withMode(CreateMode.EPHEMERAL_SEQUENTIAL).
                    forPath(Netty 服务.SERVER_PATH
                    + "/" + netAddress.getHostAddress()
                    + "#" + Constants.port + "#" + Constants.weight + "#");
        } catch (Exception e) {
            e.printStackTrace();
        }
    }else{
        //当其他事件发生时, 只需设置监听即可
        CuratorFramework client = ZookeeperFactory.create();
        client.getChildren().usingWatcher(this).forPath(Netty 服务.SERVER_PATH);
    }
}
}
}
```

分布式 RPC 服务端的修改基本上已完成, 需要注意的是, 一定要把服务端与 Zookeeper 的监听加上, 否则会出现服务端正常运行, 但 Zookeeper 中没有此服务端注册的临时节点的情况。Zookeeper 的地址, 以及 Netty 服务启动监听端口、权重、服务注册到 Zookeeper 的路径都需要写入配置文件中, 不能写死在代码里, 配置信息在第 8 章会进行修改。

客户端的修改相对服务端的修改稍微复杂些, 客户端需要新增两个类, 其中一个类主要用于管理与服务端创建的连接, 另一个类主要用于监听 Zookeeper 的 SERVER_PATH 路径的变化、获取最新的服务器列表信息、修改与服务器的连接列表。

监听类 ServerChangeWatcher 发现服务器列表, 同时创建所有服务端的连接, 当获取不到连接时, 还需提供初始化连接方法。具体实现代码如下:

```java
package com.itjoin.pro_netty.zookeeper;
import java.util.ArrayList;
import java.util.List;
import com.itjoin.pro_netty.client.ChannelFutureManager;
import com.itjoin.pro_netty.client.NettyClient;
import com.itjoin.pro_netty.constant.Constants;
import org.apache.curator.framework.CuratorFramework;
import org.apache.curator.framework.api.CuratorWatcher;
import org.apache.zookeeper.WatchedEvent;
import io.netty.channel.ChannelFuture;
public class ServerChangeWatcher implements CuratorWatcher {
    public static ServerChangeWatcher serverChangeWatcher = null;
    //客户端与服务端的连接个数需要放入配置文件中，方便调整
    //SERVER_COUNT×服务端权重=连接数
    //若单机有10000个连接，则将服务端权重设置为100
    public static final int SERVER_COUNT=100;
    public static ServerChangeWatcher getInstance(){
        if(serverChangeWatcher==null){
            serverChangeWatcher=new ServerChangeWatcher();
        }
        return serverChangeWatcher;
    }
    /**
     * 监听Zookeeper路径的变化
     * 只要有事件发生，就需要重新获取所有服务器列表并更新连接列表
     * @param event
     * @throws Exception
     */
    @Override
    public synchronized void process(WatchedEvent event) throws Exception {
        //若与Zookeeper的连接中断，则需要重连
        if (event.getState().equals(Watcher.Event.KeeperState.Disconnected)
            //连接过期，client失效
            || event.getState().equals(Watcher.Event.KeeperState.Expired)) {
            CuratorFramework client = ZookeeperFactory.recreate();
            client.getChildren().usingWatcher(this).forPath(Netty服务.SERVER_PATH);
            return ;
```

```java
    }else if(event.getState().//重新连上,并未过期
            equals(Watcher.Event.KeeperState.SyncConnected)
            && !event.equals(Watcher.Event.EventType.NodeChildrenChanged)){
CuratorFramework client = ZookeeperFactory.create();
client.getChildren().usingWatcher(this).forPath(Netty服务.SERVER_PATH);
return ;
}
System.out.println("==========重新初始化服务器连接process======");
CuratorFramework client = ZookeeperFactory.create();
//每次只能监听一次,因此需要重复加入
client.getChildren().usingWatcher(this).forPath(path);
List<String> serverPaths = client.getChildren().forPath(path);
List<String> servers = new ArrayList<>();
//获取服务器列表,并交给ChannelFutureManager保存
for(String serverPath : serverPaths){
    String[] str = serverPath.split("#");
    int weight = Integer.valueOf(str[2]);
        //不同服务器的权重值可能不同
        //此处对权重进行了简单的处理,当权重为1时,构建SERVER_COUNT个连接
        //当权重为2时,构建的连接数翻倍,依次类推
    if(weight>0){
        for(int w=0;w<=weight*SERVER_COUNT;w++){
            servers.add(str[0]+"#"+str[1]);
        }
    }
}
ChannelFutureManager.serverList.clear();
ChannelFutureManager.serverList.addAll(servers);
//根据服务器地址和IP构建连接,并交给ChannelFutureManager保存
List<ChannelFuture> futures = new ArrayList<>();
for(String realServer : ChannelFutureManager.serverList){
    String[] str = realServer.split("#");
    try {
            //此处NettyClient的bootstrap不能静态化
            ChannelFuture channelFuture = NettyClient.getBootstrap()
                .connect(str[0], Integer.valueOf(str[1])).sync();
            futures.add(channelFuture);;
```

```java
            } catch (Exception e) {
                e.printStackTrace();
            }
        }
        //加上锁，防止获取不到ChannelFuture
        synchronized (ChannelFutureManager.position){
            //先清空ChannelFuture列表
            ChannelFutureManager.clear();
            ChannelFutureManager.addAll(futures);
        }
    }
    /**
     * 初始化服务器连接列表
     * @throws Exception
     */
    public static void initChannelFuture() throws Exception {
        CuratorFramework client = ZookeeperFactory.create();
        List<String> servers = client.getChildren().forPath(Constants.SERVER_PATH);
        System.out.println("==========初始化服务器连接======");
        for(String server : servers){
            String[] str = server.split("#");
            try {
                int weight = Integer.valueOf(str[2]);
                if(weight>=0){
                    for(int w=0;w<=weight*SERVER_COUNT;w++){
                        ChannelFuture channelFuture = NettyClient.
                            getBootstrap().connect(str[0],
                            Integer.valueOf(str[1])).sync();
                        ChannelFutureManager.add(channelFuture);;
                    }
                }
            } catch (Exception e) {
                e.printStackTrace();
            }
        }
//初始化后，加上监听
client.getChildren().usingWatcher(getInstance()).forPath(Constants.SERVER_PATH);
```

```
    }
}
```

需要为客户端与服务端列表连接的链路管理类 ChannelFutureManager 提供一个线程安全的链路缓存链表，同时提供轮询获取连接链路的方法，以及其他修改缓存链表的辅助方法。具体代码如下：

```
package com.itjoin.pro_netty.client;
import java.util.concurrent.CopyOnWriteArrayList;
import java.util.concurrent.atomic.AtomicInteger;
import io.netty.channel.ChannelFuture;
public class ChannelFutureManager {
    //服务器地址列表，由于List涉及多线程的读/写
    //主要是当Zookeeper的监听类监听到服务器列表发生变化时会去修改，因此不能使用ArrayList
    static CopyOnWriteArrayList<String> serverList
        = new CopyOnWriteArrayList<String>();
    //此属性主要用来记录每次在服务器列表中获取服务器时的下标
    static AtomicInteger position=new AtomicInteger(0);
    //与服务器构建连接的ChannelFuture列表也涉及多线程的读/写
    public static CopyOnWriteArrayList<ChannelFuture> channelFutures
        = new CopyOnWriteArrayList<>();
    /**
     * 从ChannelFutureManager中获取ChannelFuture
     * 若未获取到，则通过初始化Zookeeper注册的服务器列表再次去获取
     */
    public static ChannelFuture get() throws Exception {
        ChannelFuture channelFuture = get(position);
        if(channelFuture==null){
            //初始化Zookeeper注册的服务器列表
            //在应用刚刚启动时可能会被调用
            ServerChangeWatcher.initChannelFuture();
        }
        return get(position);
    }
    /**
     * 从channelFutures中获取channelFuture
```

```
 * @param i
 * @return
 */
private static  ChannelFuture get(AtomicInteger i) {
 int size = channelFutures.size();
 if(size==0){
    return null;
 }
 ChannelFuture channel = null;
 //需要加锁,与ServerChangeWatcher类的process()方法中的锁相同
 //在获取channel时,不能清空channel链表
   synchronized(i) {
      //当下标为列表大小时需要变回0
      if (i.get() >= size) {
         channel = channelFutures.get(0);
         i.set(0);
      } else {
         //每次获取完后,下标加1
         channel = channelFutures.get(i.getAndIncrement());
      }
      //若当前get的channel不可用,则需要移除,并再次获取
      if (!channel.channel().isActive()) {
         channelFutures.remove(channel);
         return get(position);
      }
   }
   return channel;
}
     public static void removeChannel(ChannelFuture channel){
      channelFutures.remove(channel);
   }
   public static void add(ChannelFuture channel){
      channelFutures.add(channel);
   }
   public static void clear(){
         //关闭连接
         for(ChannelFuture future : channelFutures){
```

```
            future.channel().close();
    }
    channelFutures.clear();
  }
}
```

  Netty 客户端在发送数据前需要把创建与服务端连接的方式改成从 ChannelFutureManager 类中获取，最后还需要一个常量，表示 Netty 服务已注册到 Zookeeper 上的父目录。在第 8 章还会增加几个常量，整个 RPC 的服务注册和服务发现已经处理完了。还有部分参数需要根据压测情况进行相应的调整。Netty 客户端改造后的代码如下：

```java
package com.itjoin.pro_netty.client;
import java.util.concurrent.ExecutionException;
import com.alibaba.fastjson.JSONObject;
import com.itjoin.pro_netty.asyn.RequestFuture;
import io.netty.bootstrap.Bootstrap;
import io.netty.buffer.PooledByteBufAllocator;
import io.netty.channel.ChannelFuture;
import io.netty.channel.ChannelInitializer;
import io.netty.channel.ChannelOption;
import io.netty.channel.EventLoopGroup;
import io.netty.channel.nio.NioEventLoopGroup;
import io.netty.channel.socket.nio.NioSocketChannel;
import io.netty.handler.codec.LengthFieldBasedFrameDecoder;
import io.netty.handler.codec.LengthFieldPrepender;
import io.netty.handler.codec.string.StringDecoder;
import io.netty.handler.codec.string.StringEncoder;
public class NettyClient {
  //开启一个线程组，线程组一定要注意静态化
  public static EventLoopGroup group = new NioEventLoopGroup();
  public static Bootstrap getBootstrap(){
    //客户端启动辅助类
    Bootstrap boostrap = new Bootstrap();
    //设置 Socket 通道
    boostrap.channel(NioSocketChannel.class);
    boostrap.group(group);
```

```
    //设置内存分配器
    boostrap.option(ChannelOption.ALLOCATOR,PooledByteBufAllocator.DEFAULT);
    final ClientHandler handler = new ClientHandler();
    boostrap .handler(new ChannelInitializer<NioSocketChannel>()
{
    @Override
    protected void initChannel(NioSocketChannel ch)
        throws Exception {
            ch.pipeline().addLast(new
LengthFieldBasedFrameDecoder(Integer.MAX_VALUE, 0, 4, 0, 4));
        //把接收到的 ByteBuf 数据包转换成 String
        ch.pipeline().addLast(new StringDecoder());
        //业务逻辑处理 Handler
        ch.pipeline().addLast(handler);
        ch.pipeline().addLast(new LengthFieldPrepender(4, false));
        //把字符串消息转换成 ByteBuf
        ch.pipeline().addLast(new StringEncoder(Charset.forName("utf-8")));
    }
});
    return boostrap;
}
public Object sendRequest(Object msg,String path) throws
 Exception {
    try {
        //构建 request
        RequestFuture request = new RequestFuture();
        request.setPath(path);
        //设置请求 id,此处请求 id 可以设置为自动自增模式。可以采用
        //AtomicLong 类的 incrementAndGet()方法
        //此处在 RequestFuture 类里已改成了自增 id
        request.setId(1);
        //请求消息内容,此处内容可以是任意的 Java 对象
        request.setRequest(msg);
        //转换成 JSON 并发送给编码器 StringEncode
        //StringEncode 编码器再发送给 LengthFieldPrepender 长度编码器
        //最终写到 TCP 缓存中并传送给客户端
```

```
        String requestStr = JSONObject.toJSONString(request);
        ChannelFuture future = ChannelFutureManager.get();
        future.channel().writeAndFlush(requestStr);
        //同步等待响应结果,只有当 promise 有值时才会继续向下执行
        Object result = request.get();
        return result;
    } catch (Exception e) {
        e.printStackTrace();
        throw e;
    }
}
//注意:当加上动态代理并修改客户端与服务端协议后,这段代码将无法运行成功,需要修改
public static void main(String[] args) throws Exception {
    NettyClient client = new NettyClient();
    for(int i=0;i<100;i++) {
        Object result =
         client.sendRequest("id"+i,"getUserNameById");
        System.out.println(result);
    }
  }
}
public class Constants {
    public static final String SERVER_PATH = "/netty";
}
```

3.3.2 动态代理

分布式 RPC 框架在每次发送请求时都需要查看接口文档,根据接口文档封装好请求参数,指定具体的调用方法并设置请求路径,只有这样才能实现远程调用。显然,这种方式开发效率低、使用不够便捷。想要把分布式 RPC 改造成类似 Dubbo 的框架,在只引入接口的 jar 包依赖、调用对应的接口方法的情况下完成远程调用,就需要引入动态代理,整体设计如图 3-4 所示。

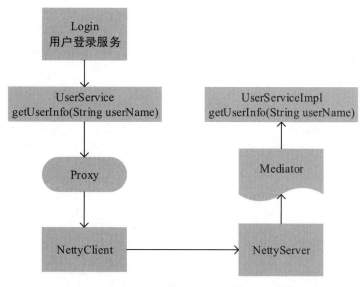

图 3-4 RPC 动态代理的调用示例

图 3-4 为用户在登录时使用动态代理获取用户信息的远程调用过程，当调用登录接口时，运行 UserService 对象的 getUserInfo()方法获取用户信息。然而，UserService 是个接口，需要在服务启动时采用 Java 的反射机制把 UserService 属性换成一个代理对象。常用的代理方式有以下两种。

（1）JDK 动态代理：通过 java.lang.reflect.Proxy 类的 newProxyInstance()方法反射生成代理类。

（2）Cglib 动态代理：采用 Enhancer 类的 create()方法生成代理类，底层利用 ASM 字节码生成框架，在内存中生成一个需要被代理类的子类。

上面这两种代理方式都可以在 RPC 中应用。在编码前先了解 Spring 的一个扩展接口——BeanPostProcessor。为了实现此接口及其两个方法，可以在 Bean 初始化前后进行一些额外的处理，这两个方法分别是 postProcessBeforeInitialization()和 postProcessAfterInitialization()。当 bean 在初始化并检测到远程调用的 Service 属性时，把对应的 Service 属性通过 Java 反射和动态代理修改成代理类，接下来分别采用两套动态代理机制完成整个代码的改造。

先使用 JDK 动态代理编写代理类 JdkProxy，其核心功能有：动态代理远程调用 RPC 服务端；在 Bean 初始化前获取所有远程调用属性并生成代理属性，以替换远程调用属性；使用注解 RemoteInvoke 标识远程调用属性。同时，由于 RequestFuture 对象在动态代理类中已构建了，所以 NettyClient 类还需要重载一个 sendRequest()方法。具体代码如下：

```java
package com.itjoin.pro_netty.proxy;
import com.alibaba.fastjson.JSONObject;
import com.itjoin.pro_netty.annotation.RemoteInvoke;
import com.itjoin.pro_netty.asyn.RequestFuture;
import com.itjoin.pro_netty.client.NettyClient;
import org.springframework.beans.BeansException;
import org.springframework.beans.factory.config.BeanPostProcessor;
import org.springframework.stereotype.Component;
import java.lang.reflect.Field;
import java.lang.reflect.InvocationHandler;
import java.lang.reflect.Method;
import java.lang.reflect.Proxy;
@Component
public class JdkProxy implements InvocationHandler,
    BeanPostProcessor {
    private Field target;
    @Override
    public Object invoke(Object proxy, Method method, Object[] args)
    throws Throwable {
        //采用 Netty 客户端调用服务端
        RequestFuture request = new RequestFuture();
        //把接口类名+方法名组装成 path
        request.setPath(target.getType().getName() + "." + method.getName());
        //设置参数（此处可以把参数改成数组）
        request.setRequest(args[0]);
        //远程调用
        Object resp = NettyClient.sendRequest(request);
        Class returnType = method.getReturnType();
        if (resp == null) {
            return null;
        }
```

```java
    //对返回结果进行反序列化处理
    resp = JSONObject.parseObject(JSONObject.toJSONString(resp),
 returnType);
        return resp;
}
//定义获取代理对象的方法
private Object getJDKProxy(Field field) {
        this.target = field;
        //JDK 动态代理只能针对接口进行代理
        return Proxy.newProxyInstance(field.getType().
getClassLoader(), new Class[]{field.getType()}, this);
}
/**
 * 在所有bean初始化完成前为包含有@RemoteInvoke注解的属性重新赋值
 */
@Override
public Object postProcessBeforeInitialization(Object bean,
String beanName) throws BeansException {
        Field[] fields = bean.getClass().getDeclaredFields();
        for (Field field : fields) {
            if (field.isAnnotationPresent(RemoteInvoke.class)) {
                field.setAccessible(true);
                try {
                    //为包含有@RemoteInvoke注解的属性重新赋值
                    field.set(bean, getJDKProxy(field));
                } catch (Exception e) {
                    e.printStackTrace();
                }
            }
        }
        return bean;
}
@Override
public Object postProcessAfterInitialization(Object bean,
String beanName) throws BeansException {
        return bean;
}
```

```
}
public static Object sendRequest(RequestFuture request) throws Exception {
    try {
        String requestStr = JSONObject.toJSONString(request);
        ChannelFuture future = ChannelFutureManager.get();
        future.channel().writeAndFlush(requestStr);
        //同步等待响应结果
        Object result = request.get();
        return result;
    } catch (Exception e) {
        e.printStackTrace();
        throw e;
    }
}
```

客户端核心代码还需要新增一个@RemoteInvoke注解：

```
package com.itjoin.pro_netty.annotation;
import java.lang.annotation.Documented;
import java.lang.annotation.ElementType;
import java.lang.annotation.Retention;
import java.lang.annotation.RetentionPolicy;
import java.lang.annotation.Target;
//此注解用于远程调用的接口属性
@Target({ElementType.FIELD})
@Retention(RetentionPolicy.RUNTIME)
@Documented
public @interface RemoteInvoke {
}
```

服务端可以去掉@Controller注解，@Remote注解可以同时作用于方法和类上，在Spring容器初始化后，不再通过@Remote注解值来作为缓存的Key，而是选择@Remote注解上的类的所有方法，通过类名+方法名的方式组装成Key，因此不再需要@Controller注解。具体代码如下：

```
//此注解既可放在方法上，又可放在类上
@Target({ElementType.TYPE,ElementType.METHOD})
@Retention(RetentionPolicy.RUNTIME)
```

```java
@Component
public @interface Remote {
    String value() default "";
}
package com.itjoin.pro_netty.spring;
import com.itjoin.pro_netty.annotation.Remote;
import com.itjoin.pro_netty.core.Mediator;
import org.springframework.context.ApplicationListener;
import org.springframework.context.event.ContextRefreshedEvent;
import org.springframework.core.Ordered;
import org.springframework.stereotype.Component;
import java.lang.reflect.Method;
import java.util.Map;
/**
 * Spring 容器初始化后，把带有@Remote 的方法与其对象加载到缓存中
 * Ordered 接口主要通过 getOrder()返回值来决定监听器运行的顺序
 * getOrder()返回值越小，运行顺序越靠前，此处为-1
 * 这是由于初始化 Remote()方法需要在 Netty 服务启动之前
 */
@Component
public class InitLoadRemoteMethod implements
     ApplicationListener<ContextRefreshedEvent> , Ordered {
    @Override
    public void onApplicationEvent(ContextRefreshedEvent
    contextRefreshedEvent) {
        //从 Spring 容器中获取标有@Remote 注解的对象
        Map<String, Object> controllerBeans =
            contextRefreshedEvent.getApplicationContext()
                .getBeansWithAnnotation(Remote.class);
        //遍历带有@Remote 注解的 bean
        for(String key : controllerBeans.keySet()){
            Object bean = controllerBeans.get(key);
            //通过反射获取 Remote 的所有方法
            Method[] methods = bean.getClass().getDeclaredMethods();
            for (Method method : methods){
                //接口名+方法名
                String methodVal = bean.getClass().
```

```
                        getInterfaces()[0].getName()
                        +"."+method.getName();
            //把方法和 bean 放入包装类 MethodBean 中
            Mediator.MethodBean methodBean =
                    new Mediator.MethodBean();
            methodBean.setBean(bean);
            methodBean.setMethod(method);
            //最终把接口名+方法名作为 Key,将方法+Bean 包装好作为 Value,放入本地缓存中
            Mediator.methodBeans.put(methodVal,methodBean);
        }
    }
}
@Override
public int getOrder() {
    return -1;
}
}
```

编写测试接口 UserService、远程实现类 UserServiceImpl、控制类 LoginController、测试类 TestProxyRpc。具体代码如下:

```
package com.itjoin.pro_netty.service;
public interface UserService {
    public Object getUserByName(String userName);
}
package com.itjoin.pro_netty.service;
import com.itjoin.pro_netty.annotation.Remote;
@Remote
public class UserServiceImpl implements UserService{
    @Override
    public Object getUserByName(String userName) {
        System.out.println("userName==="+userName);
        return "服务器响应 ok===========";
    }
}
package com.itjoin.pro_netty.controller;
import com.itjoin.pro_netty.annotation.RemoteInvoke;
import com.itjoin.pro_netty.service.UserService;
```

```java
import org.springframework.stereotype.Component;
@Component
public class LoginController {
    //调用远程服务注解
    @RemoteInvoke
    private UserService userService;
    //通过用户名称调用用户服务,以获取用户信息
    public Object getUserByName(String userName){
        return userService.getUserByName(userName);
    }
}
```

```java
package com.itjoin.pro_netty.test;
import com.itjoin.pro_netty.controller.LoginController;
import org.springframework.context.annotation.AnnotationConfigApplicationContext;
public class TestProxyRpc {
    public static void main(String[] args) {
        //注意：此处扫描包名不能包含服务器的启动类所在的包
        //只能包含具体测试类和代理类所在的包
        AnnotationConfigApplicationContext context = new
            AnnotationConfigApplicationContext(
            new String[]{"com.itjoin.pro_netty.controller",
                "com.itjoin.pro_netty.proxy"});
        LoginController loginController= context.
            getBean(LoginController.class);
        Object result = loginController.getUserByName("张三");
        System.out.println(result);
    }
}
```

如果服务器和客户端都是在本机 IDEA 启动调试的，那么一定要注意，当 Spring 容器启动扫描包名时，客户端不能包含服务器涉及的包；服务器在启动时也要修改扫描包名。启动类 ApplicationMain 的具体代码如下：

```java
package com.itjoin.pro_netty.spring;
import org.springframework.context.annotation.
 AnnotationConfigApplicationContext;
```

```java
public class ApplicationMain {
    private static volatile boolean running = true;
    public static void main(String[] args) {
        try {
            //此处扫描包名不包含代理类的包名
            //只包含service、监听器的包名
            AnnotationConfigApplicationContext context =
                    new AnnotationConfigApplicationContext(
                            "com.itjoin.pro_netty.controller",
                            "com.itjoin.pro_netty.spring"
                            ,"com.itjoin.pro_netty.service");
            //在JVM中增加一个关闭的钩子,当JVM关闭时
            // 会执行系统中已经设置的所有通过方法addShutdownHook()添加的钩子
            // 只有当系统执行完这些钩子后,JVM才会关闭
            Runtime.getRuntime().addShutdownHook(new Thread() {
                public void run() {
                    try {
                        context.stop();
                    } catch (Throwable t) {
                    }
                    synchronized (ApplicationMain.class) {
                        running = false;
                        ApplicationMain.class.notify();
                    }
                }
            });
            context.start();
        } catch (Exception e) {
            e.printStackTrace();
            System.exit(1);
        }
        System.out.println("服务器已启动====");
        synchronized (ApplicationMain.class) {
            while (running) {
                try {
                    ApplicationMain.class.wait();
                } catch (Throwable e) {
```

```
                }
            }
        }
    }
}
```

最后按顺序依次启动 Zookeeper 服务器、ApplicationMain 类，运行 TestProxyRpc 类的 main() 方法，运行结果如图 3-5 所示。

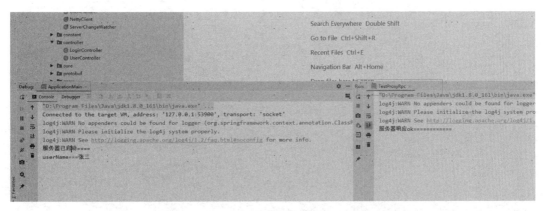

图 3-5　分布式 RPC 动态代理测试运行结果

至此，JDK 动态代理已经成功运用在分布式 RPC 服务器上了。若使用 Cglib 动态代理代替 JDK 动态代理，就需要把 JdkProxy 的 @Component 注解注释掉，同时新增一个类 CglibProxy。CglibProxy 与 JdkProxy 的区别在于动态代理属性的构建方式不同，Cglib 动态代理使用 Enhancer 代替 JDK 动态代理的 Proxy。至此，整个分布式 RPC 服务的编码全部结束了，可以对部分代码进行优化调整。例如，序列化在使用了 SPI（Service Provider Interface）技术后，可以不用写死在代码中，这样可增强代理类的扩展性。

CglibProxy 的具体代码如下：

```
package com.itjoin.pro_netty.proxy;
import com.alibaba.fastjson.JSONObject;
import com.itjoin.pro_netty.annotation.RemoteInvoke;
import com.itjoin.pro_netty.asyn.RequestFuture;
import com.itjoin.pro_netty.client.NettyClient;
```

```java
import org.springframework.beans.BeansException;
import org.springframework.beans.factory.config.BeanPostProcessor;
import org.springframework.cglib.proxy.Enhancer;
import org.springframework.cglib.proxy.MethodInterceptor;
import org.springframework.cglib.proxy.MethodProxy;
import org.springframework.stereotype.Component;
import java.lang.reflect.Field;
import java.lang.reflect.Method;
import java.util.HashMap;
import java.util.Map;
@Component
public class CglibProxy implements BeanPostProcessor{

    /**
     * 在所有 bean 初始化完成之前
     * 为 bean 中包含有 @RemoteInvoke 注解的属性重新赋值
     */
    @Override
    public Object postProcessBeforeInitialization(Object bean,
            String beanName)throws BeansException {

        Field[] fields = bean.getClass().getDeclaredFields();
        for(Field field : fields){
            if(field.isAnnotationPresent(RemoteInvoke.class)){
                field.setAccessible(true);
                final Map<Method,Class>methodClassMap=
                    new HashMap<Method,Class>();
                //此处需要把属性和属性对应的方法放入 methodClassMap 对象中
                //方便 callBack 中的 intercept()方法使用
                putMethodClass(methodClassMap,field);
                //此类是 Cglib 动态代理中非常重要的类，能动态地创建给定类的子类
                //并能拦截代理类的所有方法
                Enhancer enhancer = new Enhancer();
                //此处和 JDK 的动态代理不一样，不管是接口还是类都可以设置这个属性
                enhancer.setInterfaces(new Class[]{field.getType()});
                //设置回调方法
                enhancer.setCallback(new MethodInterceptor() {
```

```java
/**
 * 拦截代理类的所有方法
 * @param instance
 * @param method
 * @param args
 * @param proxy
 * @return
 * @throws Throwable
 */
@Override
public Object intercept(Object instance,
    Method method,Object[] args, MethodProxy proxy)
        throws Throwable {
    //采用 Netty 客户端调用服务器
    RequestFuture request = new RequestFuture();
    //把接口类名+方法名组装成 path
    request.setPath(methodClassMap.get(method).
        getName()+"."+method.getName());
    //设置参数
    request.setRequest(args[0]);
    //远程调用
    Object resp = NettyClient.sendRequest(request);
    Class returnType = method.getReturnType();
    if(resp==null){
        return null;
    }
    //对返回结果进行反序列化处理
    resp = JSONObject.parseObject(
        JSONObject.toJSONString(resp),returnType);
    return resp;
    }
});
try {
    //为包含@RemoteInvoke 注解的属性重新赋值
    field.set(bean, enhancer.create());
} catch (Exception e) {
    e.printStackTrace();
```

```
                }
            }
        }
        return bean;
    }
    /**
     * 将属性的所有方法和属性接口类型放入同一个 Map 中
     * @param methodClassMap
     * @param field
     */
    private void putMethodClass(Map<Method, Class>
                methodClassMap, Field field) {
        Method[] methods=field.getType().getDeclaredMethods();
        for(Method m : methods){
            methodClassMap.put(m, field.getType());
        }
    }
    @Override
    public Object postProcessAfterInitialization(Object bean,
                String beanName) throws BeansException {
        return bean;
    }
}
```

第 4 章

Netty 核心组件源码剖析

从本章开始对 Netty 的核心类与方法进行详细的源码剖析。反复阅读 Netty 的源码，不仅可以深入了解 Netty 的底层实现原理，对提升源码的阅读能力、自学能力也有很大的帮助。本章主要剖析与 I/O 线程模型相关的类、核心 Channel 组件、Netty 缓冲区 ByteBuf 和 Netty 内存泄漏检测组件。它们大部分分布在 io.netty.channel、io.netty.buffer、io.netty.util 等包中。NioEventLoop、AbstractChannel、AbstractByteBuf 是本章重点剖析的类。

4.1 NioEventLoopGroup 源码剖析

第 1 章中为服务启动辅助类 ServerBootstrap 设置了两个线程组，这两个线程组都是 NioEventLoopGroup 线程组。NioEventLoopGroup 线程组的功能、与 NioEventLoop 的关系及其整体设计是本节会详细剖析的内容。

NioEventLoopGroup 类主要完成以下 3 件事。

- 创建一定数量的 NioEventLoop 线程组并初始化。
- 创建线程选择器 chooser。当获取线程时，通过选择器来获取。
- 创建线程工厂并构建线程执行器。

NioEventLoopGroup 的父类为 MultithreadEventLoopGroup，父类继承了抽象类 MultithreadEventExecutorGroup。在初始化 NioEventLoopGroup 时，会调用其父类的构造方法。MultithreadEventLoopGroup 中的 DEFAULT_EVENT_LOOP_THREADS 属性决定生成多少 NioEventLoop 线程，默认为 CPU 核数的两倍，在构造方法中会把此参数传入，并最终调用 MultithreadEventExecutorGroup 类的构造方法。此构造方法运用模板设计模式来构建线程组生产模板。

线程组的生产分两步：第一步，创建一定数量的 EventExecutor 数组；第二步，通过调用子类的 newChild() 方法完成这些 EventExecutor 数组的初始化。为了提高可扩展性，Netty 的线程组除了 NioEventLoopGroup，还有 Netty 通过 JNI 方式提供的一套由 epoll 模型实现的 EpollEventLoopGroup 线程组，以及其他 I/O 多路复用模型线程组，因此 newChild() 方法由具体的线程组子类来实现。

MultithreadEventExecutorGroup 的构造方法和 NioEventLoopGroup 的 newChild() 方法的具体代码解读如下：

```
protected MultithreadEventExecutorGroup(int nThreads, Executor executor,
                                        EventExecutorChooserFactory chooserFactory,
                                        Object... args) {
    if (executor == null) {
```

```java
        //创建线程执行器及线程工厂
        executor = new ThreadPerTaskExecutor(newDefaultThreadFactory());
    }
    //根据线程数构建 EventExecutor 数组
    children = new EventExecutor[nThreads];
    for (int i = 0; i < nThreads; i ++) {
        boolean success = false;
        try {
            //初始化线程组中的线程,由 NioEventLoopGroup 创建 NioEventLoop 类实例
            children[i] = newChild(executor, args);
            success = true;
        } catch (Exception e) {
            throw new IllegalStateException("failed to create a
                                   child event loop", e);
        } finally {
            //当初始化失败时,需要优雅关闭,清理资源
            if (!success) {
                for (int j = 0; j < i; j ++) {
                    children[j].shutdownGracefully();
                }
                for (int j = 0; j < i; j ++) {
                    EventExecutor e = children[j];
                    try {
                        //当线程没有终止时,等待终止
                        while (!e.isTerminated()) {
                            e.awaitTermination(Integer.MAX_VALUE,
                            TimeUnit.SECONDS);
                        }
                    } catch (InterruptedException interrupted) {
                        Thread.currentThread().interrupt();
                        break;
                    }
                }
            }
        }
    }
//根据线程数创建选择器,选择器主要适用于 next() 方法
```

```
        chooser = chooserFactory.newChooser(children);
        final FutureListener<Object> terminationListener =
        new FutureListener<Object>(){
            @Override
            public void operationComplete(Future<Object> future) throws Exception {
                if (terminatedChildren.incrementAndGet() == children.length) {
                    terminationFuture.setSuccess(null);
                }
            }
        }
        //为每个EventLoop线程添加线程终止监听器
        for (EventExecutor e: children) {
            e.terminationFuture().addListener(terminationListener);
        }
        Set<EventExecutor> childrenSet
            = new LinkedHashSet<EventExecutor>(children.length);
        Collections.addAll(childrenSet, children);
        //创建执行器数组只读副本, 在迭代查询时使用
        readonlyChildren = Collections.unmodifiableSet(childrenSet);
    }

    @Override
    protected EventLoop newChild(Executor executor, Object... args)
        throws Exception {
        EventLoopTaskQueueFactory queueFactory = args.length == 4 ?
                            (EventLoopTaskQueueFactory) args[3] : null;
        return new NioEventLoop(this, executor, (SelectorProvider) args[0],
                ((SelectStrategyFactory) args[1]).newSelectStrategy(),
                (RejectedExecutionHandler) args[2], queueFactory);
    }
```

在newChild()方法中,NioEventLoop的初始化参数有6个:第1个参数为NioEventLoopGroup线程组本身;第 2 个参数为线程执行器,用于启动线程,在 SingleThreadEventExecutor 的 doStartThread()方法中被调用;第 3 个参数为 NIO 的 Selector 选择器的提供者;第 4 个参数主要在 NioEventLoop 的 run()方法中用于控制选择循环;第 5 个参数为非 I/O 任务提交被拒绝时的处理 Handler;第 6 个参数为队列工厂,在 NioEventLoop 中,队列读是单线程操作,而

队列写则可能是多线程操作，使用支持多生产者、单消费者的队列比较合适，默认为 MpscChunkedArrayQueue 队列。

NioEventLoopGroup 通过 next() 方法获取 NioEventLoop 线程，最终会调用其父类 MultithreadEventExecutorGroup 的 next() 方法，委托父类的选择器 EventExecutorChooser。具体使用哪种选择器对象取决于 MultithreadEventExecutorGroup 的构造方法中使用的策略模式。

根据线程条数是否为 2 的幂次来选择策略，若是，则选择器为 PowerOfTwoEventExecutorChooser，其选择策略使用与运算计算下一个选择的线程组的下标 index，此计算方法在第 7 章中也有相似的应用；若不是，则选择器为 GenericEventExecutorChooser，其选择策略为使用求余的方法计算下一个线程在线程组中的下标 index。其中，PowerOfTwoEventExecutorChooser 选择器的与运算性能会更好。

由于 Netty 的 NioEventLoop 线程被包装成了 FastThreadLocalThread 线程，同时，NioEventLoop 线程的状态由它自身管理，因此每个 NioEventLoop 线程都需要有一个线程执行器，并且在线程执行前需要通过线程工厂 io.netty.util.concurrent.DefaultThreadFactory 将其包装成 FastThreadLocalThread 线程。线程执行器 ThreadPerTaskExecutor 与 DefaultThreadFactory 的 newThread() 方法的代码解读如下：

```java
public void execute(Runnable command) {
    //调用线程工厂类的 newThread() 方法包装线程并启动
    threadFactory.newThread(command).start();
}

@Override
public Thread newThread(Runnable r) {
    //包装 FastThreadLocalThread 线程，线程名字的前缀为 NioEventLoopGroup-
    //服务启动后可通过 eclipse 或 arthas 等工具查看
    Thread t = new Thread(FastThreadLocalRunnable.wrap(r), prefix
                                    + nextId.incrementAndGet());
    try {
        if (t.isDaemon() != daemon) {
            t.setDaemon(daemon);
        }
```

```
            if (t.getPriority() != priority) {
                t.setPriority(priority);
            }
        } catch (Exception ignored) {
        }
        return t;
    }
    protected Thread newThread(Runnable r, String name) {
        return new FastThreadLocalThread(threadGroup, r, name);
    }
}
```

4.2 NioEventLoop 源码剖析

NioEventLoop 源码比 NioEventLoopGroup 源码复杂得多，每个 NioEventLoop 对象都与 NIO 中的多路复用器 Selector 一样，要管理成千上万条链路，所有链路数据的读/写事件都由它来发起。本节通过 NioEventLoop 的功能、底层设计等对其源码进行深度剖析。

NioEventLoop 有以下 5 个核心功能。

- 开启 Selector 并初始化。
- 把 ServerSocketChannel 注册到 Selector 上。
- 处理各种 I/O 事件，如 OP_ACCEPT、OP_CONNECT、OP_READ、OP_WRITE 事件。
- 执行定时调度任务。
- 解决 JDK 空轮询 bug。

NioEventLoop 这些功能的具体实现大部分都是委托其他类来完成的，其本身只完成数据流的接入工作。这种设计减轻了 NioEventLoop 的负担，同时增强了其扩展性。NioEventLoop 的整体功能如图 4-1 所示。

在图 4-1 中，第二层为 NioEventLoop 的 4 个核心方法。对于每条 EventLoop 线程来说，由于链路注册到 Selector 上的具体实现都是委托给 Unsafe 类来完成的，因此 register()方法存在于其父类 SingleThreadEventLoop 中。接下来对一些关键方法一一进行解读。

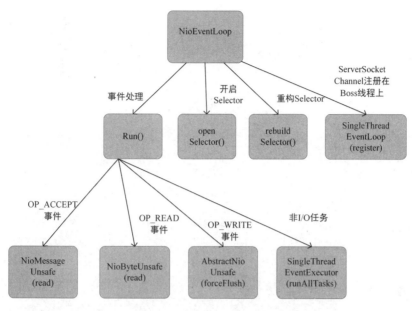

图 4-1　NioEventLoop 的整体功能

4.2.1　NioEventLoop 开启 Selector

当初始化 NioEventLoop 时，通过 openSelector() 方法开启 Selector。在 rebuildSelector() 方法中也可调用 openSelector() 方法。

在 NIO 中开启 Selector（1 行代码），只需调用 Selector.open() 或 SelectorProvider 的 openSelector() 方法即可。Netty 为 Selector 设置了优化开关，如果开启优化开关，则通过反射加载 sun.nio.ch.SelectorImpl 对象，并通过已经优化过的 SelectedSelectionKeySet 替换 sun.nio.ch.SelectorImpl 对象中的 selectedKeys 和 publicSelectedKeys 两个 HashSet 集合。其中，selectedKeys 为就绪 Key 的集合，拥有所有操作事件准备就绪的选择 Key；publicSelectedKeys 为外部访问就绪 Key 的集合代理，由 selectedKeys 集合包装成不可修改的集合。

SelectedSelectionKeySet 具体做了什么优化呢？主要是数据结构改变了，用数组替代了 HashSet，重写了 add() 和 iterator() 方法，使数组的遍历效率更高。开启优化开关，需要将系统属性 io.netty.noKeySetOptimization 设置为 true。开启 Selector 的代码解读如下：

```java
private SelectorTuple openSelector() {
    final Selector unwrappedSelector;
    try {
        //创建 Selector
        unwrappedSelector = provider.openSelector();
    } catch (IOException e) {
        throw new ChannelException("failed to open a new selector", e);
    }
    // 判断是否开启优化开关，默认没有开启直接返回 Selector
    if (DISABLE_KEY_SET_OPTIMIZATION) {
        return new SelectorTuple(unwrappedSelector);
    }
    // 通过反射创建 SelectorImpl 对象
    Object maybeSelectorImplClass = AccessController.doPrivileged(
        new PrivilegedAction<Object>() {
        @Override
        public Object run() {
            try {
                return Class.forName( "sun.nio.ch.SelectorImpl", false,
                            PlatformDependent.getSystemClassLoader());
            } catch (Throwable cause) {
                return cause;
            }
        }
    });
    // 省略了判断代码
    final Class<?> selectorImplClass = (Class<?>) maybeSelectorImplClass;
    //使用优化后的 SelectedSelectionKeySet 对象
    //将 JDK 的 sun.nio.ch.SelectorImpl.selectedKeys 替换掉
    final SelectedSelectionKeySet selectedKeySet
        = new SelectedSelectionKeySet();
    Object maybeException = AccessController.doPrivileged(
        new PrivilegedAction<Object>() {
        @Override
        public Object run() {
            try {
                Field selectedKeysField
```

```
                = selectorImplClass.getDeclaredField("selectedKeys");
            Field publicSelectedKeysField =
                selectorImplClass.getDeclaredField("publicSelectedKeys");
            //设置为可写
            Throwable cause =
                ReflectionUtil.trySetAccessible(selectedKeysField, true);
            cause =
   ReflectionUtil.trySetAccessible(publicSelectedKeysField, true);
            // 省略代码
            //通过反射的方式把 selector 的 selectedKeys 和 publicSelectedKeys
            //使用 Netty 构造的 selectedKeys 替换 JDK 的 selectedKeySet
            selectedKeysField.set(unwrappedSelector, selectedKeySet);
            publicSelectedKeysField.set
                (unwrappedSelector, selectedKeySet);
            return null;
        } catch (NoSuchFieldException e) {
            return e;
        } catch (IllegalAccessException e) {
            return e;
        }
    }
});
//把 selectedKeySet 赋给 NioEventLoop 的属性,并返回 Selector 元数据
selectedKeys = selectedKeySet;
return new SelectorTuple(unwrappedSelector,
 new SelectedSelectionKeySetSelector(unwrappedSelector, selectedKeySet));
}
```

4.2.2 NioEventLoop 的 run()方法解读

run()方法主要分三部分:select(boolean oldWakenUp),用来轮询就绪的 Channel;process SelectedKeys,用来处理轮询到的 SelectionKey;runAllTasks,用来执行队列任务。

第一部分,select(boolean oldWakenUp):主要目的是轮询看看是否有准备就绪的 Channel。在轮询过程中会调用 NIO Selector 的 selectNow()和 select(timeoutMillis)方法。由于对这两个方法的调用进行了很明显的区分,因此调用这两个方法的条件也有所不同,具体逻辑如下:

（1）当定时任务需要触发且之前未轮询过时，会调用 selectNow()方法立刻返回。

（2）当定时任务需要触发且之前轮询过（空轮询或阻塞超时轮询）直接返回时，没必要再调用 selectNow()方法。

（3）若 taskQueue 队列中有任务，且从 EventLoop 线程进入 select()方法开始后，一直无其他线程触发唤醒动作，则需要调用 selectNow()方法，并立刻返回。因为在运行 select(boolean oldWakenUp)之前，若有线程触发了 wakeUp 动作，则需要保证 tsakQueue 队列中的任务得到了及时处理，防止等待 timeoutMillis 超时后处理。

（4）当 select(timeoutMillis)阻塞运行时，在以下 4 种情况下会正常唤醒线程：其他线程执行了 wakeUp 唤醒动作、检测到就绪 Key、遇上空轮询、超时自动醒来。唤醒线程后，除了空轮询会继续轮询，其他正常情况会跳出循环。具体代码解读如下：

```
private void select(boolean oldWakenUp) throws IOException {
    Selector selector = this.selector;
    try {
        int selectCnt = 0;
        //获取当前系统的时间（纳秒级）
        long currentTimeNanos = System.nanoTime();
        //获取定时任务的触发时间
        long selectDeadLineNanos
            = currentTimeNanos + delayNanos(currentTimeNanos);
        //死循环
        for (;;) {
            //获取距离定时任务触发时间的时长（四舍五入）
            long timeoutMillis
                = (selectDeadLineNanos - currentTimeNanos + 500000L) / 1000000L;
            //已触发或超时
            if (timeoutMillis <= 0) {
                //若之前未执行过select，则调用非阻塞的selectNow()方法
                if (selectCnt == 0) {
                    selector.selectNow();
                    selectCnt = 1;
                }
```

```java
    //跳出循环，去处理 I/O 事件和定时任务
    break;
}
/**
 *当任务队列中有任务，且预唤醒标志为 false 时，需要调用 selectNow()方法
 *否则任务得不到及时处理，可能需要阻塞等待超时
 *这段判断在 Netty4 之后才加上，检测到有任务，并未设置预唤醒标识
 */
if (hasTasks() && wakenUp.compareAndSet(false, true)) {
    selector.selectNow();
    selectCnt = 1;
    break;
}
//阻塞检测就绪 Channel，除非有就绪 Channel
// 或遇空轮询问题，或者被其他线程唤醒
// 否则只能等 timeoutMillis 后自动醒来
int selectedKeys = selector.select(timeoutMillis);
//检测次数加 1，此参数主要用来判断是否为空轮询
selectCnt ++;
//若轮询到 selectKeys 不为 0，或 oldWakenUp 参数为 true
//或有线程设置 wakenUp 为 true，或任务队列和定时任务队列有值
if (selectedKeys != 0 || oldWakenUp || wakenUp.get() || hasTasks()
    || hasScheduledTasks()) {
    break;
}
//线程中断
if (Thread.interrupted()) {
    selectCnt = 1;
    break;
}
long time = System.nanoTime();
//超时自动醒来的，说明定时任务已从队列中移除了
if (time - TimeUnit.MILLISECONDS.toNanos(timeoutMillis) >=
    currentTimeNanos) {
    //将 selectCnt 设为 1，在下次进入循环时直接跳出，无须调用
    selector.selectNow()
    selectCnt = 1;
```

```
            } else if (
                //在 timeoutMillis 时间内,连续 select 次数大于或等于 512 次并未跳出循环
                SELECTOR_AUTO_REBUILD_THRESHOLD > 0 &&
                selectCnt >= SELECTOR_AUTO_REBUILD_THRESHOLD) {
                //此时进入空轮询,需要重新构建 Selector,并跳出循环
                selector = selectRebuildSelector(selectCnt);
                selectCnt = 1;
                break;
            }
            //当前系统时间更新(纳秒级)
            currentTimeNanos = time;
        }
    } catch (CancelledKeyException e) {
    }
}
```

由于上述代码逻辑过于复杂,所以绘制了如图 4-2 所示的思维导图,以便读者理解。

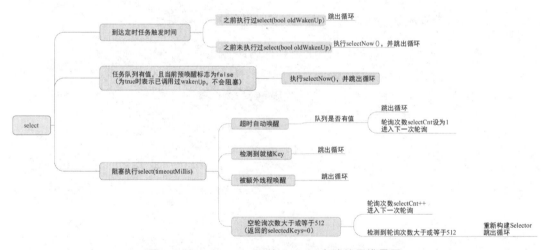

图 4-2　NioEventLoop 的 select()方法的思维导图

第二部分,processSelectedKeys:主要处理第一部分轮询到的就绪 Key,并取出这些 SelectionKey 及其附件 attachment。附件有两种类型:第一种是 AbstractNioChannel,第二种是 NioTask。其中,第二种附件在 Netty 内部未使用,因此只分析 AbstractNioChannel。根据 Key 的事件类型触发 AbstractNioChannel 的 unsafe()的不同方法。这些方法主要是 I/O 的读/写操作,

其具体源码包括附件的注册，在剖析 Channel 源码时会详细讲解。processSelectedKeys 的核心代码解读如下：

```java
private void processSelectedKeys() {
    //判断优化后的 selectedKeys 是否为空
    if (selectedKeys != null) {
        //优化处理
        processSelectedKeysOptimized();
    } else {
        //原始处理
        processSelectedKeysPlain(selector.selectedKeys());
    }
}
private void processSelectedKeysOptimized() {
    for (int i = 0; i < selectedKeys.size; ++i) {
        final SelectionKey k = selectedKeys.keys[i];
        //将 selectedKeys.keys[i]置为 null,并快速被 JVM 回收
        //无须等到调用其重置再去回收,因为 Key 的 attachment 比较大,很容易造成内存泄漏
        selectedKeys.keys[i] = null;
        final Object a = k.attachment();
        if (a instanceof AbstractNioChannel) {
            //根据 Key 的就绪事件触发对应的事件方法
            processSelectedKey(k,(AbstractNioChannel) a);
        } else {
            @SuppressWarnings("unchecked")
            NioTask<SelectableChannel> task = (NioTask<SelectableChannel>) a;
            processSelectedKey(k, task);
        }
        /**
         * 判断是否应该再次轮询
         * 每当 256 个 Channel 从 Selector 上移除时
         * 就标记 needsToSelectAgain 为 true
         */
        if (needsToSelectAgain) {
            //清空 i+1 之后的 selectedKeys
            selectedKeys.reset(i+1);
            //重新调用 selectNow()方法
```

```
            selectAgain();
            //-1+1=0,从 0 开始遍历
            i = -1;
        }
    }
}
void cancel(SelectionKey key) {
    key.cancel();
    cancelledKeys ++;
    //当移除次数大于 256 时
    if (cancelledKeys >= CLEANUP_INTERVAL) {
        cancelledKeys = 0;
        needsToSelectAgain = true;
    }
}
```

第三部分，runAllTasks：主要目的是执行 taskQueue 队列和定时任务队列中的任务，如心跳检测、异步写操作等。首先 NioEventLoop 会根据 ioRatio（I/O 事件与 taskQueue 运行的时间占比）计算任务执行时长。由于一个 NioEventLoop 线程需要管理很多 Channel，这些 Channel 的任务可能非常多，若要都执行完，则 I/O 事件可能得不到及时处理，因此每执行 64 个任务后就会检测执行任务的时间是否已用完，如果执行任务的时间用完了，就不再执行后续的任务了。具体代码解析如下：

```
protected boolean runAllTasks(long timeoutNanos) {
    //从定时任务队列中将达到执行时间的 task 丢到 taskQueue 队列中
    fetchFromScheduledTaskQueue();
    //从 taskQueue 队列获取 task
    Runnable task = pollTask();
    //若 task 为空
    if (task == null) {
        //执行 tailTasks 中的 task, 做收尾工作
        afterRunningAllTasks();
        return false;
    }
    //获取执行截止时间
    final long deadline = ScheduledFutureTask.nanoTime() + timeoutNanos;
```

```
    //执行的任务个数
    long runTasks = 0;
    //运行 task 的最后时间
    long lastExecutionTime;
    //死循环
    for (;;) {
        //运行 task 的 run()方法
        safeExecute(task);
        runTasks ++;
        //每运行 64 个任务就进行一次是否到达截止时间的检查
        if ((runTasks & 0x3F) == 0) {
            lastExecutionTime = ScheduledFutureTask.nanoTime();
            if (lastExecutionTime >= deadline) {
                break;
            }
        }
        //再从 taskQueue 队列中获取 task
        task = pollTask();
        //若没有 task 了,则更新最后执行的时间,并跳出循环
        if (task == null) {
            lastExecutionTime = ScheduledFutureTask.nanoTime();
            break;
        }
    }
    //做收尾工作
    afterRunningAllTasks();
    this.lastExecutionTime = lastExecutionTime;
    return true;
}
```

最后回到 NioEventLoop 的 run()方法,将前面的三个部分结合起来:首先调用 select(boolean oldWakenUp)方法轮询就绪的 Channel;然后调用 processSelectedKeys()方法处理 I/O 事件;最后运行 runAllTasks()方法处理任务队列。具体实现代码如下:

```
protected void run() {
    for (;;) {
        try {
```

```
try {
    /**
     * 根据是否有任务获取策略，默认策略，当有任务时，返回 selector.selectNow()
     * 当无任务时，返回 SelectStrategy.SELECT
     */
    switch (selectStrategy.calculateStrategy(selectNowSupplier,
        hasTasks())) {
        case SelectStrategy.CONTINUE:
            continue;
        case SelectStrategy.BUSY_WAIT:
        case SelectStrategy.SELECT:
            //执行 select()方法
            select(wakenUp.getAndSet(false));
            /**
             * 这里注释很长，代码后有详细讲解
             */
            if (wakenUp.get()) {
                selector.wakeup();
            }
        default:
    }
} catch (IOException e) {
    //当出现 I/O 异常时需要重新构建 Selector
    rebuildSelector0();
    handleLoopException(e);
    continue;
}
cancelledKeys = 0;
needsToSelectAgain = false;
final int ioRatio = this.ioRatio;
if (ioRatio == 100) {
    try {
        //I/O 操作，根据 selectedKeys 进行处理
        processSelectedKeys();
    } finally {
        //执行完所有任务
        runAllTasks();
```

```
            }
        } else {
            final long ioStartTime = System.nanoTime();
            try {
                //I/O 操作,根据 selectedKeys 进行处理
                processSelectedKeys();
            } finally {
                final long ioTime = System.nanoTime() - ioStartTime;
                //按一定的比例执行任务,有可能遗留一部分任务等待下次执行
                runAllTasks(ioTime * (100 - ioRatio) / ioRatio);
            }
        }
    } catch (Throwable t) {
    }
}
```

在上述代码中,有个地方很难理解,执行完 select(wakenUp.getAndSet(false))后,为何还要判断 wakenUp.get(),并去执行唤醒操作 selector.wakeup()?这个唤醒操作的底层原理是构造一个感兴趣就绪事件,让 Selector 在调用 select()方法时,轮询到就绪事件并立刻返回。WindowsSelectorImpl 的 wakeup 源码如下:

```
public Selector wakeup() {
    synchronized(this.interruptLock) {
        if (!this.interruptTriggered) {
            this.setWakeupSocket();
            this.interruptTriggered = true;
        }
        return this;
    }
}
```

由上述代码可以发现,多次同时调用 wakeup()方法与调用一次没有区别,因为 interruptTriggered 第一次调用后就为 true,后续再调用会立刻返回。在默认情况下,其他线程添加任务到 taskQueue 队列中后,会调用 NioEventLoop 的 wakeup()方法:

```
protected void wakeup(boolean inEventLoop) {
    if (!inEventLoop && wakenUp.compareAndSet(false, true)) {
        selector.wakeup();
    }
}
```

这段代码表示在变量 wakenUp 为 false 的情况下，会触发 Selector 的 wakeup 操作。再思考，若在添加任务时成功触发唤醒，那么为何 NioEventLoop 在调用 select() 方法后还要再次调用 wakenUp 呢？这段代码源码注释非常长，有点难以理解，具体如下。

（1）wakenUp 唤醒动作可能在 NioEventLoop 线程运行的两个阶段被触发，第一阶段有可能在 NioEventLoop 线程运行于 wakenUp.getAndSet(false) 与 selector.select(timeoutMillis) 之间。此时 selector.select 能立刻返回，最新任务得到及时执行。

第二阶段可能在 selector.select(timeoutMillis) 与 runAllTasks 之间，此时在 runAllTasks 执行完本次任务后又添加了新的任务，这些任务是无法被及时唤醒的。因为此时 wakenUp 为 true，其他的唤醒操作都会失败，从而导致这部分任务需要等待 select 超时后才会被执行。这对于实时性要求很高的程序来讲是无法接受的。因此在 selector.select(timeoutMillis) 与 runAllTasks 中间加入了 if(wakenUp.get())，即若有唤醒动作，则预唤醒一次，以防后续的唤醒操作失败。

（2）但由于本书的 Netty 版本，在 select() 方法里，调用 hasTask() 查看任务队列是否有任务。且在进入 select() 方法前，会把 wakenUp 设置为 false，所以 wakenUp.compareAndSet(false, true) 会成功。因此，当添加了新的任务时会调用 selectNow() 方法，不会等到超时才执行任务。因此无须在 select() 方法后再次调用 wakeup() 方法。

（3）wakeup() 方法操作耗性能，因此建议在非复杂处理时，尽量不开额外线程。

4.2.3　NioEventLoop 重新构建 Selector 和 Channel 的注册

从 select 函数的代码解读中发现，Netty 在空轮询次数大于或等于阈值（默认 512）时，需要重新构建 Selector。重新构建 Selector 的方式比较巧妙：重新打开一个新的 Selector，将旧的 Selector 上的 key 和 attchment 复制过去，同时关闭旧的 Selector。具体代码解读如下：

```java
private void rebuildSelector0() {
    final Selector oldSelector = selector;
    final NioEventLoop.SelectorTuple newSelectorTuple;
    if (oldSelector == null) {
        return;
    }
    try {
        //开启新的 Selector
        newSelectorTuple = openSelector();
    } catch (Exception e) {
        return;
    }
    int nChannels = 0;
    //遍历旧的 Selector 上的 key
    for (SelectionKey key: oldSelector.keys()) {
        Object a = key.attachment();
        try {
            //判断 key 是否有效
            if (!key.isValid() ||
       key.channel().keyFor(newSelectorTuple.unwrappedSelector) != null) {
                continue;
            }
            //在旧的 Selector 上触发的事件需要取消
            int interestOps = key.interestOps();
            key.cancel();
            //把 Channel 重新注册到新的 Selector 上
            SelectionKey newKey
                = key.channel().register(newSelectorTuple.unwrappedSelector,
                                         interestOps, a);
            if (a instanceof AbstractNioChannel) {
                ((AbstractNioChannel) a).selectionKey = newKey;
            }
            nChannels ++;
        } catch (Exception e) {
        }
    }
    selector = newSelectorTuple.selector;
```

```
        unwrappedSelector = newSelectorTuple.unwrappedSelector;
        try {
            //关闭旧的 Selector
            oldSelector.close();
        } catch (Throwable t) {
        }
    }
```

注册方法 register() 在两个地方被调用：一是在端口绑定前，需要把 NioServerSocketChannel 注册到 Boss 线程的 Selector 上；二是当 NioEventLoop 监听到有链路接入时，把链路 SocketChannel 包装成 NioSocketChannel，并注册到 Woker 线程中。最终调用 NioSocketChannel 的辅助对象 unsafe 的 register() 方法，unsafe 执行父类 AbstractUnsafe 的 register() 模板方法（在第 5 章会进行详细剖析）。

4.3　Channel 源码剖析

Channel 是 Netty 抽象出来的对网络 I/O 进行读/写的相关接口，与 NIO 中的 Channel 接口相似。Channel 的主要功能有网络 I/O 的读/写、客户端发起连接、主动关闭连接、关闭链路、获取通信双方的网络地址等。Channel 接口下有一个重要的抽象类——AbstractChannel，一些公共的基础方法都在这个抽象类中实现，一些特定功能可以通过各个不同的实现类去实现。最大限度地实现了功能和接口的重用。

AbstractChannel 融合了 Netty 的线程模型、事件驱动模型，但由于网络 I/O 模型及协议种类比较多，除了 TCP 协议，Netty 还支持很多其他连接协议，并且每种协议都有传统阻塞 I/O 和 NIO（非阻塞 I/O）版本的区别。不同协议、不同阻塞类型的连接有不同的 Channel 类型与之对应，因此 AbstractChannel 并没有与网络 I/O 直接相关的操作。每种阻塞与非阻塞 Channel 在 AbstractChannel 上都会继续抽象一层，如 AbstractNioChannel，既是 Netty 重新封装的 Epoll SocketChannel 实现，其他非阻塞 I/O Channel 的抽象层，接下来分别对这些核心 Channel 进行详细讲解。

4.3.1 AbstractChannel 源码剖析

AbstractChannel 抽象类包含以下几个重要属性。

- EventLoop：每个 Channel 对应一条 EventLoop 线程。
- DefaultChannelPipeline：一个 Handler 的容器，也可以将其理解为一个 Handler 链。Handler 主要处理数据的编/解码和业务逻辑。
- Unsafe：实现具体的连接与读/写数据，如网络的读/写、链路关闭、发起连接等。命名为 Unsafe 表示不对外提供使用，并非不安全。

图 4-3 为 AbstractChannel 的功能图，从图中可以看出，Unsafe 属性的功能非常丰富，在 AbstractChannel 类中有一个 Unsafe 抽象类——AbstractUnsafe，其具体的实现类在 AbstractChannel 的子类中，AbstractUnsafe 的大部分方法都采用了模板设计模式，具体的实现细节由其子类完成。例如，bind()方法：

```
public final void bind(final SocketAddress localAddress, final ChannelPromise promise) {
    boolean wasActive = isActive();
    try {
        //模板设计模式：调用子类 NioServerSocketChannel 的 doBind()方法
        doBind(localAddress);
    } catch (Throwable t) {
        //绑定失败回调
        safeSetFailure(promise, t);
        closeIfClosed();
        return;
    }
    //从非活跃状态到活跃状态触发了 active 事件
    if (!wasActive && isActive()) {
        invokeLater(new Runnable() {
            @Override
            public void run() {
                pipeline.fireChannelActive();
            }
        });
```

```
        }
        //绑定成功回调通知
        safeSetSuccess(promise);
    }
```

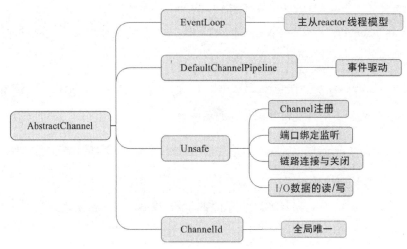

图 4-3　AbstractChannel 的功能图

4.3.2　AbstractNioChannel 源码剖析

AbstractNioChannel 也是一个抽象类，不过它在 AbstractChannel 的基础上增加了一些属性和方法，AbstractChannel 没有涉及 NIO 的任何属性和具体方法，包括 AbstractUnsafe。AbstractNioChannel 有以下 3 个重要属性：

```
private final SelectableChannel ch;//真正用到的 NIO Channel
protected final int readInterestOp;//监听感兴趣的事件
volatile SelectionKey selectionKey;//注册到 Selector 后获取的 key
```

SelectableChannel 是 java.nio.SocketChannel 和 java.nio.ServerSocketChannel 公共的抽象类；readInterestOp 用于区分当前 Channel 监听的事件类型；至于 selectionKey，它是将 SelectableChannel 注册到 Selector 后的返回值。从这些属性的定义可以看出，在 AbstractNioChannel 中，已经将 Netty 的 Channel 和 Java NIO 的 Channel 关联起来了。

AbstractNioChannel 的方法都很简洁，下面只解读其中一个很重要的方法——doRegister()：

```java
/**
 * doRegister()方法在 AbstractUnsafe 的 register0()方法中被调用
 * @throws Exception
 */
protected void doRegister() throws Exception {
    boolean selected = false;
    for (;;) {
        try {
            /**
             * 通过 javaChannel()方法获取具体的 Nio Channel
             * 把 Channel 注册到其 EventLoop 线程的 Selector 上
             * 对于注册后返回的 selectionKey，需要为其设置 Channel 感兴趣的事件
             */
            selectionKey
                = javaChannel().register(eventLoop().unwrappedSelector(),
                                        0, this);
            return;
        } catch (CancelledKeyException e) {
            if (!selected) {
                //由于尚未调用 select.select(…)
                //因此可能仍在缓存而未删除但已取消 selectionKey
                //强制调用 selector.selectNow()方法
                //将已经取消的 selectionKey 从 Selector 上删除
                eventLoop().selectNow();
                selected = true;
            } else {
                //只有第一次抛出此异常，才能调用 selector.selectNow()进行取消
                //如果调用 selector.selectNow() 还有取消的缓存，则可能是 JDK 的一个 bug
                throw e;
            }
        }
    }
}
```

doRegister()方法在 AbstractUnsafe 的 register0()方法中被调用。在 AbstractNioChannel 中有个非常重要的类——AbstractNioUnsafe，是 AbstractUnsafe 类的 NIO 实现，主要实现了 connect()、flush0()等方法。它还实现了 NioUnsafe 接口，实现了其 finishConnect()、forceFlush()、ch()等

方法。其中，forceFlush()与flush0()最终调用的NioSocketChannel的doWrite()方法来完成缓存数据写入Socket的工作，在剖析NioSocketChannel时会详细讲解。connect()和finishConnect()这两个方法只有在Netty客户端中才会使用到。下面是对这两个方法进行的详细分析。

connect()方法代码解读：

```
@Override
public final void connect(
        final SocketAddress remoteAddress,
        final SocketAddress localAddress,
        final ChannelPromise promise) {
    //设置任务为不可取消的状态，并确定Channel已打开
    if (!promise.setUncancellable() || !ensureOpen(promise)) {
        return;
    }
    try {
        //确保没有正在进行的连接
        if (connectPromise != null) {
            throw new ConnectionPendingException();
        }
        //获取之前的状态
        boolean wasActive = isActive();
        //在远程连接时，会出现以下3种结果
        // (1) 连接成功，返回true
        // (2) 暂时没有连接上，服务端没有返回ACK应答，连接结果不确定，返回false
        // (3) 连接失败，直接抛出I/O异常
        // 由于协议和I/O模型不同，连接的方式也不一致，因此具体实现由其子类完成
        if (doConnect(remoteAddress, localAddress)) {
            //连接成功后会触发ChannelActive事件
            // 最终会将NioSocketChannel中的selectionKey
            // 设置为SelectionKey.OP_READ
            // 用于监听网络读操作位
            fulfillConnectPromise(promise, wasActive);
        } else {
            connectPromise = promise;
            requestedRemoteAddress = remoteAddress;
```

```java
//获取连接超时时间
int connectTimeoutMillis = config().getConnectTimeoutMillis();
if (connectTimeoutMillis > 0) {
    //根据连接超时时间设置定时任务
    connectTimeoutFuture = eventLoop().schedule(new Runnable() {
        @Override
        public void run() {
            //到达超时时间后触发校验
            ChannelPromise connectPromise
                = AbstractNioChannel.this.connectPromise;
            ConnectTimeoutException cause
                = new ConnectTimeoutException(
                    "connection timed out: " + remoteAddress);
            if (connectPromise != null &&
                connectPromise.tryFailure(cause)) {
                //如果发现连接并没有完成，则关闭连接句柄，释放资源
                //设置异常堆栈并发起取消注册操作
                close(voidPromise());
            }
        }
    }, connectTimeoutMillis, TimeUnit.MILLISECONDS);
}
//增加连接结果监听器
promise.addListener(new ChannelFutureListener() {
    @Override
    public void operationComplete(ChannelFuture future) throws
     Exception {
        //如果接收到连接完成的通知，则判断连接是否被取消
        //如果被取消则关闭连接句柄，释放资源，发起取消注册操作
        if (future.isCancelled()) {
            if (connectTimeoutFuture != null) {
                connectTimeoutFuture.cancel(false);
            }
            connectPromise = null;
            close(voidPromise());
        }
```

```
        });
    }
} catch (Throwable t) {
    promise.tryFailure(annotateConnectException(t, remoteAddress));
    //关闭连接句柄，释放资源，发起取消注册操作
    //从多路复用器上移除
    closeIfClosed();
}
```

finishConnect()方法代码解读：

```
@Override
public final void finishConnect() {
    //只有EventLoop线程才能调用finishConnect()方法
    //此方法在NioEventLoop的processSelectedKey()方法中被调用
    assert eventLoop().inEventLoop();
    try {
        boolean wasActive = isActive();
        /**
         * 判断连接结果（由其子类完成）
         * 通过SocketChannel的finishConnect()方法判断连接结果
         * 连接成功返回true
         * 连接失败抛异常
         * 链路被关闭、链路中断等异常也属于连接失败
         */
        doFinishConnect();
        //负责将SocketChannel修改为监听读操作位
        //用来监听网络的读事件
        fulfillConnectPromise(connectPromise, wasActive);
    } catch (Throwable t) {
        //连接失败，关闭连接句柄，释放资源，发起取消注册操作
        fulfillConnectPromise(connectPromise,annotateConnectException(t,
                    requestedRemoteAddress));
    } finally {
        if (connectTimeoutFuture != null) {
            connectTimeoutFuture.cancel(false);
        }
```

```
        connectPromise = null;
    }
}
```

4.3.3　AbstractNioByteChannel 源码剖析

AbstractNioChannel 拥有 NIO 的 Channel, 具备 NIO 的注册、连接等功能。但 I/O 的读/写交给了其子类, Netty 对 I/O 的读/写分为 POJO 对象与 ByteBuf 和 FileRegion, 因此在 AbstractNioChannel 的基础上继续抽象了一层, 分为 AbstractNioMessageChannel 与 AbstractNioByteChannel。本小节详细讲解 AbstractNioByteChannel。它发送和读取的对象是 ByteBuf 与 FileRegion 类型。图 4-4 展示了 AbstractNioByteChannel 的基本功能和属性。

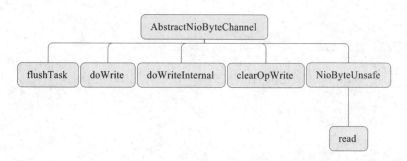

图 4-4　AbstractNioByteChannel 的基本功能和属性

属性 flushTask 为 task 任务, 主要负责刷新发送缓存链表中的数据, 由于 write 的数据没有直接写在 Socket 中, 而是写在了 ChannelOutboundBuffer 缓存中, 所以当调用 flush() 方法时, 会把数据写入 Socket 中并向网络中发送。因此当缓存中的数据未发送完成时, 需要将此任务添加到 EventLoop 线程中, 等待 EventLoop 线程的再次发送。

doWrite() 与 doWriteInternal() 方法在 AbstractChannel 的 flush0() 方法中被调用, 主要功能是从 ChannelOutboundBuffer 缓存中获取待发送的数据, 进行循环发送, 发送的结果分为以下 3 种。

(1) 发送成功, 跳出循环直接返回。

(2) 由于 TCP 缓存区已满, 成功发送的字节数为 0, 跳出循环, 并将写操作 OP_WRITE 事件添加到选择 Key 兴趣事件集中。

（3）默认当写了 16 次数据还未发送完时，把选择 Key 的 OP_WRITE 事件从兴趣事件集中移除，并添加一个 flushTask 任务，先去执行其他任务，当检测到此任务时再发送。具体代码解析如下：

```
protected void doWrite(ChannelOutboundBuffer in) throws Exception {
    //写请求自循环次数，默认为16次
    int writeSpinCount = config().getWriteSpinCount();
    do {
        //获取当前 Channel 的缓存 ChannelOutboundBuffer 中的当前待刷新消息
        Object msg = in.current();
        //所有消息都发送成功了
        if (msg == null) {
            //清除 Channel 选择 Key 兴趣事件集中的 OP_WRITE 写操作事件
            clearOpWrite();
            //直接返回，没必要再添加写任务
            return;
        }
        //发送数据
        writeSpinCount -= doWriteInternal(in, msg);
    } while (writeSpinCount > 0);
    /**
     * 当因缓冲区满了而发送失败时
     * doWriteInternal 返回 Integer.MAX_VALUE
     * 此时 writeSpinCount<0 为 true
     * 当发送 16 次还未全部发送完，但每次都写成功时
     * writeSpinCount 为 0
     */
    incompleteWrite(writeSpinCount < 0);
}

protected final void incompleteWrite(boolean setOpWrite) {
    if (setOpWrite) {
        //将 OP_WRITE 写操作事件添加到 Channel 的选择 Key 兴趣事件集中
        setOpWrite();
    } else {
        //清除 Channel 选择 Key 兴趣事件集中的 OP_WRITE 写操作事件
        clearOpWrite();
        //将写操作任务添加到 EventLoop 线程上，以便后续继续发送
```

```java
            eventLoop().execute(flushTask);
        }
    }
}
private int doWriteInternal(ChannelOutboundBuffer in, Object msg) throws
 Exception {
        if (msg instanceof ByteBuf) {
            ByteBuf buf = (ByteBuf) msg;
            if (!buf.isReadable()) {
                //若可读字节数为0，则从缓存区中移除
                in.remove();
                return 0;
            }
            //实际发送字节数据
            final int localFlushedAmount = doWriteBytes(buf);
            if (localFlushedAmount > 0) {
                //更新字节数据的发送进度
                in.progress(localFlushedAmount);
                if (!buf.isReadable()) {
                    //若可读字节数为0，则从缓存区中移除
                    in.remove();
                }
                return 1;
            }
        } else if (msg instanceof FileRegion) {
            //如果是文件FileRegion消息
            FileRegion region = (FileRegion) msg;
            if (region.transferred() >= region.count()) {
                in.remove();
                return 0;
            }
            //实际写操作
            long localFlushedAmount = doWriteFileRegion(region);
            if (localFlushedAmount > 0) {
                //更新数据的发送进度
                in.progress(localFlushedAmount);
                if (region.transferred() >= region.count()) {
                    //若region已全部发送成功，
```

```
                //则从缓存中移除
                in.remove();
            }
            return 1;
        }
    } else {
        //不支持发送其他类型的数据
        throw new Error();
    }
    //当实际发送字节数为 0 时,返回 Integer.MAX_VALUE
    return WRITE_STATUS_SNDBUF_FULL;
}
```

NioByteUnsafe 的 read()方法的实现思路大概分为以下 3 步。

（1）获取 Channel 的配置对象、内存分配器 ByteBufAllocator，并计算内存分配器 RecvByteBufAllocator.Handle。

（2）进入 for 循环。循环体的作用：使用内存分配器获取数据容器 ByteBuf，调用 doReadBytes()方法将数据读取到容器中，如果本次循环没有读到数据或链路已关闭，则跳出循环。另外，当循环次数达到属性 METADATA 的 defaultMaxMessagesPerRead 次数（默认为 16）时，也会跳出循环。由于 TCP 传输会产生粘包问题，因此每次读取都会触发 channelRead 事件，进而调用业务逻辑处理 Handler。

（3）跳出循环后，表示本次读取已完成。调用 allocHandle 的 readComplete()方法，并记录读取记录，用于下次分配合理内存。

具体代码解析如下：

```
public final void read() {
    //获取 pipeline 通道配置、Channel 管道
    final ChannelConfig config = config();
    //socketChannel 已关闭
    if (shouldBreakReadReady(config)) {
        clearReadPending();
        return;
```

```java
}
final ChannelPipeline pipeline = pipeline();
//获取内存分配器,默认为PooledByteBufAllocator
final ByteBufAllocator allocator = config.getAllocator();
final RecvByteBufAllocator.Handle allocHandle = recvBufAllocHandle();
//清空上一次读取的字节数,每次读取时均重新计算
//字节buf分配器,并计算字节buf分配器Handler
allocHandle.reset(config);
ByteBuf byteBuf = null;
boolean close = false;
try {
    do {
        //分配内存
        byteBuf = allocHandle.allocate(allocator);
        //读取通道接收缓冲区的数据
        allocHandle.lastBytesRead(doReadBytes(byteBuf));
        if (allocHandle.lastBytesRead() <= 0) {
            //若没有数据可读取,则释放内存
            byteBuf.release();
            byteBuf = null;
            close = allocHandle.lastBytesRead() < 0;
            if (close) {
                // 当读到-1时,表示Channel通道已关闭
                // 没必要再继续读
                readPending = false;
            }
            break;
        }
        //更新读取消息计数器
        allocHandle.incMessagesRead(1);
        readPending = false;
        //通知通道处理读取数据,触发Channel管道的fireChannelRead事件
        pipeline.fireChannelRead(byteBuf);
        byteBuf = null;
    } while (allocHandle.continueReading());
    //读取操作完毕
    allocHandle.readComplete();
```

```
            //触发 Channel 管道的 fireChannelReadComplete 事件
            pipeline.fireChannelReadComplete();
            if (close) {
                //如果 Socket 通道关闭,则关闭读操作
                closeOnRead(pipeline);
            }
        } catch (Throwable t) {
            //处理读异常
            handleReadException(pipeline, byteBuf, t, close, allocHandle);
        } finally {
            if (!readPending && !config.isAutoRead()) {
                //若读操作完毕,且没有配置自动读
                //则从选择 Key 兴趣集中移除读操作事件
                removeReadOp();
            }
        }
    }
}
```

4.3.4　AbstractNioMessageChannel 源码剖析

AbstractNioMessageChannel 写入和读取的数据类型是 Object,而不是字节流,那么它的读/写方法与 AbstractNioByteChannel 的读/写方法有哪些不同呢?下面进行详细讲解。

在读数据时,AbstractNioMessageChannel 数据不存在粘包问题,因此 AbstractNioMessageChannel 在 read()方法中先循环读取数据包,再触发 channelRead 事件。

在写数据时,AbstractNioMessageChannel 数据逻辑简单。它把缓存 outboundBuffer 中的数据包依次写入 Channel 中。如果 Channel 写满了,或循环写、默认写的次数为子类 Channel 属性 METADATA 中的 defaultMaxMessagesPerRead 次数,则在 Channel 的 SelectionKey 上设置 OP_WRITE 事件,随后退出,其后 OP_WRITE 事件处理逻辑和 Byte 字节流写逻辑一样。read() 与 doWrite()方法的代码解读如下:

```
public void read() {
    assert eventLoop().inEventLoop();
```

```java
//获取 Channel 的配置对象
final ChannelConfig config = config();
final ChannelPipeline pipeline = pipeline();
//获取计算内存分配 Handle
final RecvByteBufAllocator.Handle allocHandle
    = unsafe().recvBufAllocHandle();
//清空上次的记录
allocHandle.reset(config);
boolean closed = false;
Throwable exception = null;
try {
    try {
        do {
            /**
             * 调用子类的 doReadMessages()方法
             * 读取数据包,并放入 readBuf 链表中
             * 当成功读取时返回 1
             */
            int localRead = doReadMessages(readBuf);
            //已无数据,跳出循环
            if (localRead == 0) {
                break;
            }
            //链路关闭,跳出循环
            if (localRead < 0) {
                closed = true;
                break;
            }
            //记录成功读取的次数
            allocHandle.incMessagesRead(localRead);
            //默认不能超过 16 次
        } while (allocHandle.continueReading());
    } catch (Throwable t) {
        exception = t;
    }
    int size = readBuf.size();
    //循环处理读取的数据包
```

```java
            for (int i = 0; i < size; i ++) {
                readPending = false;
                //触发 channelRead 事件
                pipeline.fireChannelRead(readBuf.get(i));
            }
            readBuf.clear();
            //记录当前读取记录，以便下次分配合理内存
            allocHandle.readComplete();
            //触发 readComplete 事件
            pipeline.fireChannelReadComplete();
            if (exception != null) {
                //处理 Channel 异常关闭
                closed = closeOnReadError(exception);
                pipeline.fireExceptionCaught(exception);
            }
            if (closed) {
                inputShutdown = true;
                //处理 Channel 正常关闭
                if (isOpen()) {
                    close(voidPromise());
                }
            }
        } finally {
            //读操作完毕，且没有配置自动读
            if (!readPending && !config.isAutoRead()) {
                //移除读操作事件
                removeReadOp();
            }
        }
    }
}
protected void doWrite(ChannelOutboundBuffer in) throws Exception {
    final SelectionKey key = selectionKey();
    //获取 Key 兴趣集
    final int interestOps = key.interestOps();
    for (;;) {
        Object msg = in.current();
```

```java
if (msg == null) {
    //数据已全部发送完，从兴趣集中移除 OP_WRITE 事件
    if ((interestOps & SelectionKey.OP_WRITE) != 0) {
        key.interestOps(interestOps & ~SelectionKey.OP_WRITE);
    }
    break;
}
try {
    boolean done = false;
    //获取配置中循环写的最大次数
    for (int i = config().getWriteSpinCount() - 1; i >= 0; i--) {
        //调用子类方法，若 msg 写成功了，则返回 true
        if (doWriteMessage(msg, in)) {
            done = true;
            break;
        }
    }
    //若发送成功，则将其从缓存链表中移除
    //继续发送下一个缓存节点数据
    if (done) {
        in.remove();
    } else {
        //若没有写成功，则 doWriteMessage 返回 false
        if ((interestOps & SelectionKey.OP_WRITE) == 0) {
            // 将 OP_WRITE 事件添加到兴趣事件集中
            key.interestOps(interestOps | SelectionKey.OP_WRITE);
        }
        break;
    }
} catch (Exception e) {
    //当出现异常时，判断是否继续写
    if (continueOnWriteError()) {
        in.remove(e);
    } else {
        throw e;
    }
}
```

            }
        }

4.3.5 NioSocketChannel 源码剖析

之前分析的 Channel 都是抽象类，NioSocketChannel 是 AbstractNioByteChannel 的子类，也是 io.netty.channel.socket.SocketChannel 的实现类。Netty 服务的每个 Socket 连接都会生成一个 NioSocketChannel 对象。NioSocketChannel 在 AbstractNioByteChannel 的基础上封装了 NIO 中的 SocketChannel，实现了 I/O 的读/写与连接操作，其核心功能如下。

- SocketChannel 在 NioSocketChannel 构造方法中由 SelectorProvider.provider().openSocketChannel()创建，提供 javaChannel()方法以获取 SocketChannel。
- 实现 doReadBytes()方法，从 SocketChannel 中读取数据。
- 重写 doWrite()方法、实现 doWriteBytes()方法，将数据写入 Socket 中。
- 实现 doConnect()方法和客户端连接。

图 4-5 为 NioSocketChannel 的核心功能图，注明了这些功能会在哪些地方被调用，从图中可以看出，大部分方法都被其辅助对象 Unsafe 调用。

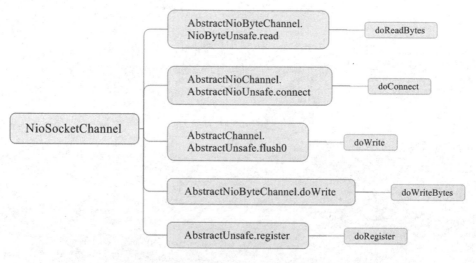

图 4-5　NioSocketChannel 的核心功能图

I/O 的读/写的核心代码解读如下：

```java
protected int doReadBytes(ByteBuf byteBuf) throws Exception {
    //获取计算内存分配器 Handle
    final RecvByteBufAllocator.Handle allocHandle =unsafe().recvBufAllocHandle();
    //设置尝试读取字节数为 buf 的可写字节数
    allocHandle.attemptedBytesRead(byteBuf.writableBytes());
    //从 Channel 中读取字节并写入 buf 中，返回读取的字节数
    return byteBuf.writeBytes(javaChannel(),allocHandle.attemptedBytesRead());
}
protected int doWriteBytes(ByteBuf buf) throws Exception {
    //获取 buf 的可读字节数
    final int expectedWrittenBytes = buf.readableBytes();
    //把 buf 写入 Socket 缓存中，返回写入的字节数
    return buf.readBytes(javaChannel(), expectedWrittenBytes);
}
protected void doWrite(ChannelOutboundBuffer in) throws Exception {
    //获取 SocketChannel
    SocketChannel ch = javaChannel();
    //获取配置属性 writeSpinCount（循环写的最大次数）
    int writeSpinCount = config().getWriteSpinCount();
    do {
        //缓存中数据为空，无数据可写
        if (in.isEmpty()) {
            //移除写事件，并直接返回
            clearOpWrite();
            return;
        }
        //获取一次最大可写字节数
        int maxBytesPerGatheringWrite =
((NioSocketChannel.NioSocketChannelConfig)config).getMaxBytesPerGatheringWrite();
        /**
         * 缓存由多个 Entry 组成，每次写时都可能写多个 Entry
         * 具体一次性该发送多少数据
         * 由 ByteBuffer 数组的最大长度和一次最大可写字节数决定
         */
        ByteBuffer[] nioBuffers = in.nioBuffers(1024,maxBytesPerGatheringWrite);
        int nioBufferCnt = in.nioBufferCount();
```

```java
//缓存中有多少个nioBuffer
switch (nioBufferCnt) {
    case 0:
        //非ByteBuffer数据,交给父类实现
        writeSpinCount -= doWrite0(in);
        break;
    case 1: {
        ByteBuffer buffer = nioBuffers[0];
        //buf可读字节数
        int attemptedBytes = buffer.remaining();
        //把buf发送到Socket缓存中
        final int localWrittenBytes = ch.write(buffer);
        //发送失败
        if (localWrittenBytes <= 0) {
            //将写事件添加到事件兴趣集中
            incompleteWrite(true);
            return;
        }
        /**
         * 根据成功写入字节数和尝试写入字节数调整下次最大可写字节数
         * 当两者相等时,若尝试写入字节数*2 大于当前最大写入字节数
         * 则下次最大可写字节数等于尝试写入字节数*2
         * 当两者不相等,成功写入字节数小于尝试写入字节数/2,
         * 且尝试写入字节数大于4096时
         * 下次最大可写字节数等于尝试写入字节数/2
         */
        adjustMaxBytesPerGatheringWrite(attemptedBytes,
                localWrittenBytes, maxBytesPerGatheringWrite);
        //从缓存中移除写入字节数
        in.removeBytes(localWrittenBytes);
        //循环写次数减1
        --writeSpinCount;
        break;
    }
    default: {
        //尝试写入字节数
        long attemptedBytes = in.nioBufferSize();
```

```
            //真正发送到 Socket 缓存中的字节数
            final long localWrittenBytes = ch.write(nioBuffers, 0,nioBufferCnt);
            //如果发送失败
            if (localWrittenBytes <= 0) {
                //将写事件添加到事件兴趣集中
                //以便下次 NioEventLoop 继续触发写操作
                incompleteWrite(true);
                return;
            }
            //调整下次最大可写字节数
            adjustMaxBytesPerGatheringWrite((int) attemptedBytes,
                    (int) localWrittenBytes,
                maxBytesPerGatheringWrite);
            //从缓存中移除发送成功的字节
            in.removeBytes(localWrittenBytes);
            //循环写次数减 1
            --writeSpinCount;
            break;
        }
    }
} while (writeSpinCount > 0);
/**
 * 未全部发送完:
 * 若 writeSpinCount<0
 * 则说明 Socket 缓冲区已满, 未发送成功
 * 若 writeSpinCount=0
 * 则说明 Netty 缓存数据太大, 写了 16 次还未写完
 */
incompleteWrite(writeSpinCount < 0);
}
```

4.3.6　NioServerSocketChannel 源码剖析

NioServerSocketChannel 是 AbstractNioMessageChannel 的子类, 由于 NioServerSocketChannel 由服务端使用, 并且只负责监听 Socket 的接入, 不关心 I/O 的读/写, 所以与 NioSocketChannel 相比要简单很多。它封装了 NIO 中的 ServerSocketChannel, 并通过 newSocket()方法打开 ServerSocket

Channel。它的多路复用器注册与 NioSocketChannel 的多路复用器注册一样,由父类 AbstractNio Channel 实现。下面重点关注它是如何监听新加入的连接的(需要由 doReadMessages()方法来完成)。具体代码解析如下:

```
protected int doReadMessages(List<Object> buf) throws Exception {
    //调用 serverSocketChannel.accept()监听新加入的连接
    SocketChannel ch = SocketUtils.accept(javaChannel());
    try {
        if (ch != null) {
            //每个新连接都会构建一个 NioSocketChannel
            buf.add(new NioSocketChannel(this, ch));
            return 1;
        }
    } catch (Throwable t) {
        try {
            //若连接出现异常,则关闭
            ch.close();
        } catch (Throwable t2) {
        }
    }
    return 0;
}
```

4.4 Netty 缓冲区 ByteBuf 源码剖析

在网络传输中,字节是基本单位,NIO 使用 ByteBuffer 作为 Byte 字节容器,但是其使用过于复杂。因此 Netty 写了一套 Channel,代替了 NIO 的 Channel。Netty 缓冲区又采用了一套 ByteBuf 代替了 NIO 的 ByteBuffer。Netty 的 ByteBuf 子类非常多,这里只对核心的 ByteBuf 进行详细的剖析。图 4-6 展示了 ByteBuf 的主要特性,图中列出的类是本节重点剖析的类。图 4-6 中的前 3 个特性对 NIO ByteBuffer 的缺点进行了改进。

第 4 章 Netty 核心组件源码剖析

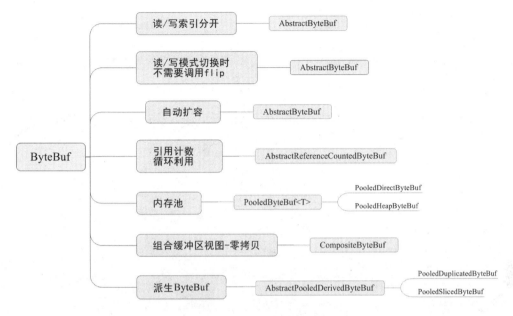

图 4-6　ByteBuf 的主要特性

NIO ByteBuffer 只有一个位置指针 position，在切换读/写状态时，需要手动调用 flip()方法或 rewind()方法，以改变 position 的值，而且 ByteBuffer 的长度是固定的，一旦分配完成就不能再进行扩容和收缩，当需要放入或存储的对象大于 ByteBuffer 的容量时会发生异常。每次编码时都要进行可写空间校验。

Netty 的 AbstractByteBuf 将读/写指针分离，同时在写操作时进行了自动扩容。对其使用而言，无须关心底层实现，且操作简便、代码无冗余。

NIO ByteBuffer 的 duplicate()方法可以复制对象，复制后的对象与原对象共享缓冲区的内存，但其位置指针独立维护。Netty 的 ByteBuf 也采用了这功能，并设计了内存池。内存池是由一定大小和数量的内存块 ByteBuf 组成的，这些内存块的大小默认为 16MB。当从 Channel 中读取数据时，无须每次都分配新的 ByteBuf，只需从大的内存块中共享一份内存，并初始化其大小及独立维护读/写指针即可。Netty 采用对象引用计数，需要手动回收。每复制一份 ByteBuf 或派生出新的 ByteBuf，其引用值都需要增加。

4.4.1　AbstractByteBuf 源码剖析

AbstractByteBuf 是 ByteBuf 的子类，它定义了一些公共属性，如读索引、写索引、mark、最大容量等。AbstractByteBuf 实现了一套读/写操作的模板方法，其缓冲区真正的数据读/写由其子类完成。图 4-7 展示了 AbstractByteBuf 的核心功能，接下来对其核心功的源码进行剖析。

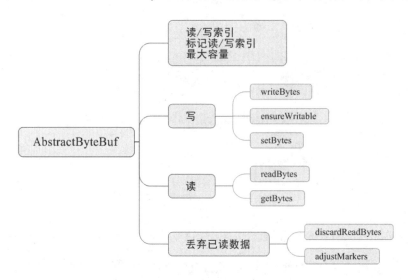

图 4-7　AbstractByteBuf 的核心功能

AbstractByteBuf 的核心属性及说明如下：

```
//读索引
int readerIndex;
//写索引
int writerIndex;
/**
 * 标记读索引
 * 在解码时，由于消息不完整，无法处理
 * 需要将 readerIndex 复位
 * 此时需要先为索引做个标记
 */
private int markedReaderIndex;
//标记写索引
private int markedWriterIndex;
```

```
//最大容量
private int maxCapacity;
```

AbstractByteBuf 的写操作 writeBytes()方法涉及扩容，在扩容时除了合法校验，还需要计算新的容量值，若内存大小为 2 的整数次幂，则 AbstractByteBuf 的子类比较好分配内存，因此扩容后的大小必须是 2 的整数次幂，计算逻辑复杂。具体代码解读如下：

```
public ByteBuf writeBytes(byte[] src, int srcIndex, int length) {
    //确保可写,当容量不足时自动扩容
    ensureWritable(length);
    //缓冲区真正的写操作由子类实现
    setBytes(writerIndex, src, srcIndex, length);
    //调整写索引
    writerIndex += length;
    return this;
}
final void ensureWritable0(int minWritableBytes) {
    //获取 ByteBuf 对象的引用计数
    //如果返回值为零,则说明该对象被销毁,会抛异常
    ensureAccessible();
    //若可写字节数大于 minWritableBytes,则无须扩容
    if (minWritableBytes <= writableBytes()) {
        return;
    }
    //获取写索引
    final int writerIndex = writerIndex();
    /**checkBounds
     * 判断将要写入的字节数是否大于最大可写字节数（maxCapacity-writerIndex）
     * 如果大于则直接抛异常,否则继续执行
     */
    if (checkBounds) {
        if (minWritableBytes > maxCapacity - writerIndex) {
            throw new IndexOutOfBoundsException(String.format(
                "writerIndex(%d) + minWritableBytes(%d) exceeds " +
                    "maxCapacity(%d): %s",
                writerIndex, minWritableBytes, maxCapacity, this));
        }
```

```java
        }
        //最小容量
        int minNewCapacity = writerIndex + minWritableBytes;
        //计算自动扩容后的容量,需满足最小容量,必须是2的幂数
        int newCapacity
            = alloc().calculateNewCapacity(minNewCapacity,maxCapacity);
        /**
         * maxFastWritableBytes 返回不用复制和重新分配内存的最快、最大可写字节数
         * 默认等于 writableBytes()
         */
        int fastCapacity = writerIndex + maxFastWritableBytes();
        //减少重新分配内存
        if (newCapacity > fastCapacity && minNewCapacity <= fastCapacity) {
            newCapacity = fastCapacity;
        }
        //由子类将容量调整到新的容量值
        capacity(newCapacity);
    }
    boolean isAccessible() {
        //引用是否大于0
        return refCnt() != 0;
    }
    /**
     * 当 threshold 小于阈值(4MB)时,新的容量(newCapacity)都是以64为基数向左移位计算出来的
     * 通过循环,每次移动1位,直到 newCapacity>=minNewCapacity
     * 如果计算出来的 newCapacity 大于 maxCapacity,则返回 maxCapacity
     * 否则返回 newCapacity
     * 当 minNewCapacity>=阈值(4MB)时
     * 先计算 minNewCapacity/threshold*threshold 的大小
     * 如果这个值加上一个 threshold(4MB)大于 newCapacity
     * 则 newCapacity 的值取 maxCapacity;
     * 否则 newCapacity=minNewCapacity/threshold*threshold+threshold
     * @param minNewCapacity
     * @param maxCapacity
     * @return
     */
    public int calculateNewCapacity(int minNewCapacity, int maxCapacity) {
```

```java
//检查 minNewCapacity 是否大于 0
checkPositiveOrZero(minNewCapacity, "minNewCapacity");
//阈值为 4MB
final int threshold = CALCULATE_THRESHOLD;
if (minNewCapacity == threshold) {
    return threshold;
}
//当大于 4MB 时
if (minNewCapacity > threshold) {
    //先获取离 minNewCapacity 最近的 4MB 的整数倍值,且小于 minNewCapacity
    int newCapacity = minNewCapacity/threshold*threshold;
    /**
     * 此处新的容量值不会倍增,因为 4MB 以上内存比较大
     * 如果继续倍增,则可能带来额外的内存浪费
     * 只能在此基础上+4MB,并判断是否大于 maxCapacity
     * 若大于则返回 maxCapacity
     * 否则返回 newCapacity+threshold
     */
    if (newCapacity > maxCapacity - threshold) {
        newCapacity = maxCapacity;
    } else {
        newCapacity += threshold;
    }
    return newCapacity;
}
/**
 * 当小于 4MB 时,以 64 为基础倍增
 * 64->128->256…直到满足最小容量要求,并以此容量值作为新容量值
 */
int newCapacity = 64;
while (newCapacity < minNewCapacity) {
    newCapacity <<= 1;
}
return Math.min(newCapacity, maxCapacity);
}
```

读操作 readBytes()方法的源码解读如下：

```
@Override
public ByteBuf readBytes(byte[] dst, int dstIndex, int length) {
    //检测 ByteBuf 是否可读
    //检测其可读长度是否小于 length
    checkReadableBytes(length);
    //数据的具体读取由子类实现
    getBytes(readerIndex, dst, dstIndex, length);
    //修改读索引
    readerIndex += length;
    return this;
}
```

readBytes()方法调用 getBytes()方法从当前的读索引开始，将 length 个字节复制到目标 byte 数组中。由于不同的子类对应不同的复制操作，所以 AbstractByteBuf 类中的 getBytes()方法是一个抽象方法，留给子类来实现。下面是一个具体的子类 PooledHeapByteBuf 对 getBytes()方法的实现代码：

```
@Override
public final ByteBuf getBytes(int index, byte[] dst, int dstIndex, int length) {
    //先检查目标数组的存储空间是否够用，再检查 ByteBuf 的可读内容是否足够
    checkDstIndex(index, length, dstIndex, dst.length);
    //将 ByteBuf 中的内容读取到 dst 数组中
    System.arraycopy(memory, idx(index), dst, dstIndex, length);
    return this;
}
```

另一个子类 PooledDirectByteBuf 对 getBytes()方法的实现代码如下：

```
@Override
public ByteBuf getBytes(int index, byte[] dst, int dstIndex, int length) {
    //检查
    checkDstIndex(index, length, dstIndex, dst.length);
    //将 NIO 的 ByteBuffer 中的内容获取到 dst 数组中
    _internalNioBuffer(index, length, true).get(dst, dstIndex, length);
    return this;
```

```
    }
    final ByteBuffer _internalNioBuffer(int index, int length,
     boolean duplicate) {
        //根据 readIndex 获取偏移量 offset
        index = idx(index);
        //从 memory 中复制一份内存对象，两者共享缓存区，但其位置指针独立维护
        ByteBuffer buffer
            = duplicate ? newInternalNioBuffer(memory) : internalNioBuffer();
        //设置新的 ByteBuffer 位置及其最大长度
        buffer.limit(index + length).position(index);
        return buffer;
    }
}
```

对 AbstractByteBuf 的核心部分已基本上了解了，下面通过对 AbstractReferenceCountedByteBuf 类进行深入剖析来了解 Netty 是如何运用引用计数法管理 ByteBuf 生命周期的。

4.4.2　AbstractReferenceCountedByteBuf 源码剖析

Netty 在进行 I/O 的读/写时使用了堆外直接内存，实现了零拷贝，堆外直接内存 Direct Buffer 的分配与回收效率要远远低于 JVM 堆内存上对象的创建与回收速率。Netty 使用引用计数法来管理 Buffer 的引用与释放。Netty 采用了内存池设计，先分配一块大内存，然后不断地重复利用这块内存。例如，当从 SocketChannel 中读取数据时，先在大内存块中切一小部分来使用，由于与大内存共享缓存区，所以需要增加大内存的引用值，当用完小内存后，再将其放回大内存块中，同时减少其引用值。

运用到引用计数法的 ByteBuf 大部分都需要继承 AbstractReferenceCountedByteBuf 类。该类有个引用值属性——refCnt，其功能大部分与此属性有关。图 4-8 为 AbstractReferenceCountedByteBuf 的功能图。

由于 ByteBuf 的操作可能存在多线程并发使用的情况，其 refCnt 属性的操作必须是线程安全的，因此采用了 volatile 来修饰，以保证其多线程可见。在 Netty 中，ByteBuf 会被大量地创建，为了节省内存开销，通过 AtomicIntegerFieldUpdater 来更新 refCnt 的值，而没有采用 AtomicInteger

类型。因为 AtomicInteger 类型创建的对象比 int 类型多占用 16B 的对象头，当有几十万或几百万 ByteBuf 对象时，节约的内存可能就是几十 MB 或几百 MB。

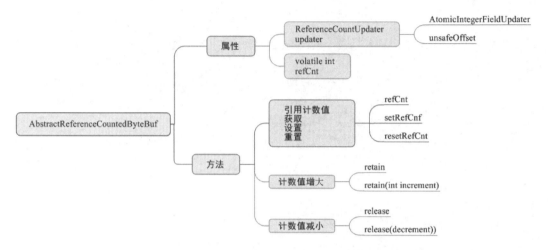

图 4-8　AbstractReferenceCountedByteBuf 的功能图

AbstractReferenceCountedByteBuf 的大部分功能都是由 updater 属性完成的，其核心属性解读如下：

```
    /**
     * 调用 Unsafe 类的 objectFieldOffset()方法
     * 以获取某个字段相对于 Java 对象的起始地址的偏移量
     * Netty 为了提升性能，构建了 Unsafe 对象
     * 采用此偏移量访问 ByteBuf 的 refCnt 字段
     * 并未直接使用 AtomicIntegerFieldUpdater 来操作
     */
    private static final long REFCNT_FIELD_OFFSET
        = ReferenceCountUpdater.getUnsafeOffset(
            AbstractReferenceCountedByteBuf.class, "refCnt");
    /**
     * AtomicIntegerFieldUpdater 属性委托给 ReferenceCountUpdater 来管理
     * 主要用于更新和获取 refCnt 的值
     */
private static final
AtomicIntegerFieldUpdater<AbstractReferenceCountedByteBuf>
```

```
AIF_UPDATER =
AtomicIntegerFieldUpdater.newUpdater(AbstractReferenceCountedByteBuf.class,
        "refCnt");
    /**
    * 引用计数值的实际管理者
    */
    private static final
ReferenceCountUpdater<AbstractReferenceCountedByteBuf> updater =
        new ReferenceCountUpdater<AbstractReferenceCountedByteBuf>() {
            @Override
            protected
              AtomicIntegerFieldUpdater<AbstractReferenceCountedByteBuf>
              updater() {
                return AIF_UPDATER;
            }
            @Override
            protected long unsafeOffset() {
                return REFCNT_FIELD_OFFSET;
            }
        }
    /**
    * 引用计数值，初始化为 2，与调用 refCnt() 获取的实际值 1 有差别
    */
    @SuppressWarnings("unused")
    private volatile int refCnt = updater.initialValue();
```

在旧的版本中，refCnt 引用计数的值每次加 1 或减 1，默认为 1，大于 0 表示可用，等于 0 表示已释放。在 Netty v4.1.38.Final 版本中，refCnt 的初始值为 2，每次操作也不同。那么，为何要改成这种设计方式？有什么好处呢？通过 4.4.3 小节来解决这个疑问。

4.4.3 ReferenceCountUpdater 源码剖析

ReferenceCountUpdater 是 AbstractReferenceCountedByteBuf 的辅助类，用于完成对引用计数值的具体操作，其功能如图 4-9 所示。虽然它的所有功能基本上都与引用计数有关，但与 Netty 之前的版本相比有很大的改动，主要是 Netty v4.1.38.Final 版本采用了乐观锁方式来修改

refCnt，并在修改后进行校验。例如，retain()方法在增加了refCnt后，如果出现了溢出，则回滚并抛异常。在旧版本中，采用的是原子性操作，不断地提前判断，并尝试调用compareAndSet。与之相比，新版本的吞吐量有所提高，但若还是采用refCnt的原有方式，从1开始每次加1或减1，则会引发一些问题，需要重新设计。这也是新版本改动较大的主要原因。

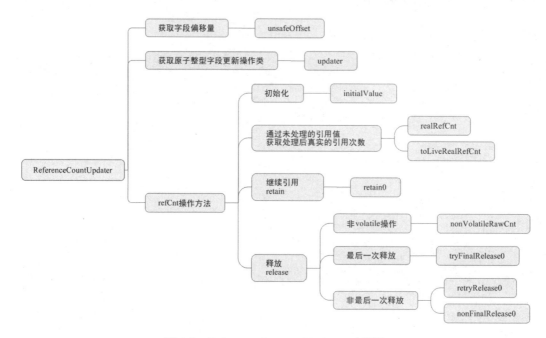

图4-9　ReferenceCountedUpdater 功能图

由 duplicate()、slice()衍生的 ByteBuf 与原对象共享底层的 Buffer，原对象的引用可能需要增加，引用增加的方法为 retain0()。retain0()方法为 retain()方法的具体实现，其代码解读如下：

```
private T retain0(T instance, final int increment, final int rawIncrement) {
    //乐观锁，先获取原值，再增加
    int oldRef = updater().getAndAdd(instance, rawIncrement);
    //如果原值不为偶数，则表示 ByteBuf 已经被释放，无法继续引用
    if (oldRef != 2 && oldRef != 4 && (oldRef & 1) != 0) {
        throw new IllegalReferenceCountException(0, increment);
    }
    // 如果增加后出现了溢出
    if ((oldRef <= 0 && oldRef + rawIncrement >= 0)
```

```
            || (oldRef >= 0 && oldRef + rawIncrement < oldRef)) {
            //则回滚并抛出异常
            updater().getAndAdd(instance, -rawIncrement);
        throw new IllegalReferenceCountException(realRefCnt(oldRef), increment);
        }
        return instance;
    }
```

旧版本代码如下：

```
private ByteBuf retain0(int increment) {
    //一直循环
    for (;;) {
        int refCnt = this.refCnt;
        final int nextCnt = refCnt + increment;
        //先判断是否溢出
        if (nextCnt <= increment) {
            throw new IllegalReferenceCountException(refCnt, increment);
        }
        //如果引用在for循环体中未被修改过，则用新的引用值替换
        if (refCntUpdater.compareAndSet(this, refCnt, nextCnt)) {
            break;
        }
    }
    return this;
}
```

在进行引用计数的修改时，并不会先判断是否会出现溢出，而是先执行，执行完之后再进行判断，如果溢出则进行回滚。在高并发情况下，与之前的版本相比，Netty v4.1.38.Final 的吞吐量会有所提升，但 refCnt 不是每次都进行加 1 或减 1 的操作，主要原因是修改前无判断。若有多条线程同时操作，则线程 1 调用 ByteBuf 的 release()方法，线程 2 调用 retain()方法，线程 3 调用 release()方法。

线程 1 执行完后，refCnt 的值为 0；线程 2 执行完 retain()方法后，正好执行完增加操作，refCnt 此时由 0 变成 1，还未执行到判断回滚环节；此时线程 3 执行 release()方法，能正常运

行，导致 ByteBuf 出现多次销毁操作。若采用奇数表示销毁状态，偶数表示正常状态，则该问题得以解决，最终释放后会变成奇数。

ByteBuf 使用完后需要执行 release() 方法。release() 方法的返回值为 true 或 false，false 表示还有引用存在；true 表示无引用，此时会调用 ByteBuf 的 deallocate() 方法进行销毁。相关代码解读如下：

```
public final boolean release(T instance) {
    /**
     * 先采用普通方法获取 refCnt 的值，无须采用 volatile 获取
     * 因为 tryFinalRelease0() 方法会用 CAS 更新
     * 若更新失败了，则通过 retryRelease0() 方法进行不断循环处理
     * 此处一开始并非调用 retryRelease0() 方法循环尝试来修改 refCnt 的值
     * 这样设计，吞吐量会有所提升
     * 当 rawCnt 不等于 2 时，说明还在其他地方引用了此对象
     * 调用 nonFinalRelease0() 方法，尝试采用 CAS 使 refCnt 的值减 2
     */
    int rawCnt = nonVolatileRawCnt(instance);
    return rawCnt == 2 ? tryFinalRelease0(instance, 2) ||
        retryRelease0(instance, 1)
        : nonFinalRelease0(instance, 1, rawCnt, toLiveRealRefCnt(rawCnt, 1));
}
/**
 * 采用 CAS 最终释放，将 refCnt 设置为 1
 */
private boolean tryFinalRelease0(T instance, int expectRawCnt) {
    return updater().compareAndSet(instance, expectRawCnt, 1);    }
private boolean retryRelease0(T instance, int decrement) {
    for (;;) {
        /**
         * volatile 获取 refCnt 的原始值
         * 并通过 toLiveRealRefCnt() 方法将其转化成真正引用次数
         * 原始值必须是 2 的倍数，否则状态为已释放，会抛异常
         */
        int rawCnt = updater().get(instance), realCnt = toLiveRealRefCnt(rawCnt, decrement);
        //如果引用次数与当前释放次数相等
```

```java
        if (decrement == realCnt) {
            /**
             * 尝试最终释放，采用 CAS 更新 refCnt 的值为 1，若更新成功则返回 true
             * 如果更新失败，说明 refCnt 的值改变了，则继续进行循环处理
             */
            if (tryFinalRelease0(instance, rawCnt)) {
                return true;
            }
        } else if (decrement < realCnt) {
            /**
             * 引用次数大于当前释放次数
             * CAS 更新 refCnt 的值
             * 引用原始值-2*当前释放次数
             * 此处释放为非最后一次释放
             * 因此释放成功后会返回 false
             */
            if (updater().compareAndSet(instance, rawCnt, rawCnt - (decrement
                << 1))) {
                return false;
            }
        } else {
            throw new IllegalReferenceCountException(realCnt, -decrement);
        }
        Thread.yield();
    }
}
/**
 * 非最后一次释放，realCnt>1
 */
private boolean nonFinalRelease0(T instance, int decrement, int rawCnt,
int realCnt) {
    //与 retryRelease0()方法中的其中一个释放分支一样
    if (decrement < realCnt
            && updater().compareAndSet(instance, rawCnt, rawCnt - (decrement
        << 1))) {
        return false;
    }
```

```
    //若CAS更新失败,则进入retryRelease0
    return retryRelease0(instance, decrement);
}
private static int toLiveRealRefCnt(int rawCnt, int decrement) {
    //当rawCnt为偶数时,真实引用值需要右移1位
    if (rawCnt == 2 || rawCnt == 4 || (rawCnt & 1) == 0) {
        return rawCnt >>> 1;
    }
    //rawCnt为奇数表示已释放,此时会抛出异常
    throw new IllegalReferenceCountException(0, -decrement);
}
private int nonVolatileRawCnt(T instance) {
    //获取偏移量
    final long offset = unsafeOffset();
    /**
     * 若偏移量正常,则选择Unsafe的普通get
     * 若偏移量获取异常,则选择Unsafe的volatile get
     */
    return offset != -1 ? PlatformDependent.getInt(instance, offset) :
            updater().get(instance);
}
```

ReferenceCountUpdater 主要运用 JDK 的 CAS 来修改计数器,为了提高性能,还引入了 Unsafe 类,可直接操作内存。至此,ByteBuf 的引用计数告一段落,下面会对 Netty 的另一种零拷贝方式(组合缓冲区视图 CompositeByteBuf)进行详细剖析。

4.4.4　CompositeByteBuf 源码剖析

CompositeByteBuf 的主要功能是组合多个 ByteBuf,对外提供统一的 readerIndex 和 writerIndex。由于它只是将多个 ByteBuf 的实例组装到一起形成了一个统一的视图,并没有对 ByteBuf 中的数据进行拷贝,因此也属于 Netty 零拷贝的一种,主要应用于编码和解码。

例如,将消息头和消息体两个 ByteBuf 组合到一块进行编码,可能会觉得 Netty 有写缓冲区,其本身就会存储多个 ByteBuf,此时只需把两个 ByteBuf 分别写入缓冲区 ChannelOutboundBuffer 即可,没必要使用组合 ByteBuf。但是在将 ByteBuf 写入缓冲区之前,需要对整个消息进行编

码，如长度编码，此时需要把两个 ByteBuf 合并成一个，无须额外处理就可以知道其整体长度。因此使用 CompositeByteBuf 是非常适合的。

在解码时，由于 Socket 通信传输数据会产生粘包和半包问题，因此需要一个读半包字节容器，这个容器采用 CompositeByteBuf 比较合适，将每次从 Socket 中读到的数据直接放入此容器中，少了一次数据的拷贝。Netty 的解码类 ByteToMessageDecoder 默认的读半包字节容器 Cumulator 未采用 CompositeByteBuf，此时可在其子类中调用 setCumulator 进行修改。但需要注意的是，CompositeByteBuf 需要依赖具体的使用场景。因为 CompositeByteBuf 使用了复杂的算法逻辑，所以其效率有可能比使用内存拷贝的低。

CompositeByteBuf 内部定义了一个 Component 类型的集合。实际上，Component 是 ByteBuf 的包装实现类，它聚合了 ByteBuf 对象并维护了 ByteBuf 对象在集合中的位置偏移量信息等。图 4-10 介绍了 CompositeByteBuf 功能。下面结合它在 ByteToMessageDecoder 解码器中的应用对其源码进行详细的解读。

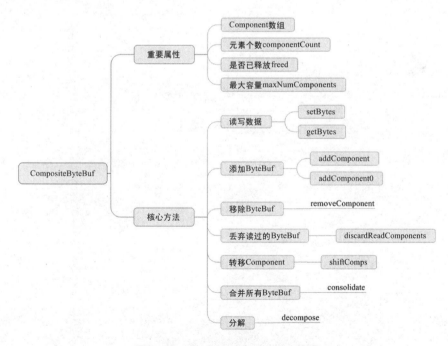

图 4-10　CompositeByteBuf 功能图

在图 4-10 中，addComponent()方法及相关方法的代码解读如下：

```java
/**
 * 添加 ByteBuf
 */
public CompositeByteBuf addComponent(boolean increaseWriterIndex, int cIndex,
 ByteBuf buffer) {
    checkNotNull(buffer, "buffer");
    //把 buffer 加入 Component 数组中
    //并对数组中的元素进行相应的挪动
    addComponent0(increaseWriterIndex, cIndex, buffer);
    //是否需要合并成一个 ByteBuf
    consolidateIfNeeded();
    return this;
}
private int addComponent0(boolean increaseWriterIndex, int cIndex,
    ByteBuf buffer) {
    assert buffer != null;
    boolean wasAdded = false;
    try {
        //检查下标是否正常
        checkComponentIndex(cIndex);
        //构建包装 component
        CompositeByteBuf.Component c = newComponent(buffer, 0);
        int readableBytes = c.length();
        //把 component 追加到数组中，并移动其后面的元素
        addComp(cIndex, c);
        wasAdded = true;
        if (readableBytes > 0 && cIndex < componentCount - 1) {
            /**
             * 当插入的位置不在数组末尾时
             * 不仅需要设置插入元素的位置信息
             * 还需要更新其后面元素的位置信息
             */
            updateComponentOffsets(cIndex);
        } else if (cIndex > 0) {
            //当插入的位置在数组末尾时，只需设置插入元素的位置信息即可
            c.reposition(components[cIndex - 1].endOffset);
```

```java
            }
            //是否修改写索引
            if (increaseWriterIndex) {
                writerIndex += readableBytes;
            }
            return cIndex;
        } finally {
            //当出现异常增加失败时,释放 buffer
            if (!wasAdded) {
                buffer.release();
            }
        }
    }
}
private CompositeByteBuf.Component newComponent(ByteBuf buf, int offset) {
    if (checkAccessible && !buf.isAccessible()) {
        throw new IllegalReferenceCountException(0);
    }
    //获取 buf 读索引及 buf 的长度
    int srcIndex = buf.readerIndex(), len = buf.readableBytes();
    ByteBuf slice = null;
    //若是派生 ByteBuf,则需要通过 unwrap 得到原始 ByteBuf
    //原始 buf 的读索引=派生 buf 读索引+偏移量 adjustment
    //由于是非可重复利用内存,所以其读索引应为 0
    if (buf instanceof AbstractUnpooledSlicedByteBuf) {
        srcIndex += ((AbstractUnpooledSlicedByteBuf) buf).idx(0);
        slice = buf;
        buf = buf.unwrap();
    } else if (buf instanceof PooledSlicedByteBuf) {
        srcIndex += ((PooledSlicedByteBuf) buf).adjustment;
        slice = buf;
        buf = buf.unwrap();
    }
    //包装成 Component 对象返回,并设置为大端模式(与网络传输模式一致)
    return new CompositeByteBuf.Component(buf.order(ByteOrder.BIG_ENDIAN),
            srcIndex, offset, len, slice);
}
private void consolidateIfNeeded() {
```

```java
        //若components数组中的元素超过了其允许的最大容量
        //则需要把所有ByteBuf合并成一个
        int size = componentCount;
        if (size > maxNumComponents) {
            final int capacity = components[size - 1].endOffset;
            ByteBuf consolidated = allocBuffer(capacity);
            lastAccessed = null;
            //循环遍历所有ByteBuf,并把数据写入consolidated中
            for (int i = 0; i < size; i ++) {
                components[i].transferTo(consolidated);
            }
            components[0]
                = new CompositeByteBuf.Component(consolidated, 0, 0, capacity, consolidated);
            //移除components数组中下标为1~(size-1)的元素
            removeCompRange(1, size);
        }
    }
private void removeCompRange(int from, int to) {
    if (from >= to) {
        return;
    }
    final int size = componentCount;
    assert from >= 0 && to <= size;
    /**
     * 若只移除中间元素
     * 则需要把后面的元素向前移
     */
    if (to < size) {
        System.arraycopy(components, to, components, from, size - to);
    }
    //处理后的元素个数
    int newSize = size - to + from;
    //把移除的元素置空
    for (int i = newSize; i < size; i++) {
        components[i] = null;
    }
    //更新元素个数
```

```
        componentCount = newSize;
    }
```

读/写数据及其相关方法解读如下：

```
/**
 * 读取数据并写到 dst 中
 * @param index      读索引
 * @param dst        目标缓存
 * @param dstIndex   目标缓存写索引
 * @param length     读取长度
 * @return
 */
@Override
public CompositeByteBuf getBytes(int index, ByteBuf dst, int dstIndex,
        int length) {
    //检查 index, length, dstIndex, dst.capacity()是否合法
    checkDstIndex(index, length, dstIndex, dst.capacity());
    if (length == 0) {
        return this;
    }
    //根据 readerIndex 获取 components 数组的下标
    int i = toComponentIndex0(index);
    /**
     * 由于 ByteBuf 是逻辑组合
     * 在读的过程中，一个 buf 可能不够
     * 需要从多个 buf 中读取数据，因此需要 while 循环，直到写满
     */
    while (length > 0) {
        CompositeByteBuf.Component c = components[i];
        //在每次读数据时，只能读取当前 buf 的可读字节与 length 两者中的最小值
        int localLength = Math.min(length, c.endOffset - index);
        //从 buf 中读取 localLength 字节到 dst 中
        c.buf.getBytes(c.idx(index), dst, dstIndex, localLength);
        //其读索引值需要增加 localLength
        index += localLength;
        //目标 buf 的写索引也需进行相应的增加
        dstIndex += localLength;
```

```java
            //对需要写的字节数进行相应的调整
            length -= localLength;
            //components数组的下标也要向上移一位
            i++;
        }
        return this;
    }
    //通过偏移量获取对应的下标
    private int toComponentIndex0(int offset) {
        int size = componentCount;
        //偏移量为0，快速获取第一个元素
        if (offset == 0) {
            for (int i = 0; i < size; i++) {
                if (components[i].endOffset > 0) {
                    return i;
                }
            }
        }
        //当少于或等于两个元素时，没必要使用二分查找算法，只需快速判断并获取即可
        if (size <= 2) {
            return size == 1 || offset < components[0].endOffset ? 0 : 1;
        }
        /**
         * 当components数组中的元素个数多于两个时，使用二分查找算法
         * 其分割规则主要根据偏移量来判断
         * 当偏移量大于或等于元素的endOffset时，low=mid+1
         * 当偏移量小于遍历元素的offset时，high=mid-1
         * 当偏移量等于遍历元素的offset时，只需返回其下标即可
         */
        for (int low = 0, high = size; low <= high;) {
            int mid = low + high >>> 1;
            CompositeByteBuf.Component c = components[mid];
            if (offset >= c.endOffset) {
                low = mid + 1;
            } else if (offset < c.offset) {
                high = mid - 1;
            } else {
```

```java
            return mid;
        }
    }
    //若没有找到，则抛异常
    throw new Error("should not reach here");
}
/**
 * 从src缓冲区读取数据并写入CompositeByteBuf中
 * @param index
 * @param src
 * @param srcIndex
 * @param length
 * @return
 */
@Override
public CompositeByteBuf setBytes(int index, byte[] src, int srcIndex,
int length) {
    //检查
    checkSrcIndex(index, length, srcIndex, src.length);
    if (length == 0) {
        return this;
    }
    //根据writerIndex获取components数组的下标
    int i = toComponentIndex0(index);
    /**
     * 循环写入，逻辑与循环读数据逻辑类似
     * 只是index从readerIndex换成了writerIndex
     */
    while (length > 0) {
        CompositeByteBuf.Component c = components[i];
        int localLength = Math.min(length, c.endOffset - index);
        c.buf.setBytes(c.idx(index), src, srcIndex, localLength);
        index += localLength;
        srcIndex += localLength;
        length -= localLength;
        i ++;
    }
```

```
        return this;
}
```

通过以上代码剖析，对 CompositeByteBuf 的底层实现有了更进一步的了解，明白了它的内部是如何处理数据的读/写、如何添加新元素的。但细心的读者会发现，虽然 Component 是 ByteBuf 的包装对象，但它并没有像其他派生对象一样调用 retain()方法。ByteBuf 的引用计数器并没有任何的改变，这个问题可以通过解读 CompositeByteBuf 在 ByteToMessageDecoder 解码器中的源码来找到答案。具体的代码解读如下：

```
//复合缓冲区实现读半包字节容器
public static final ByteToMessageDecoder.Cumulator COMPOSITE_CUMULATOR = new
    ByteToMessageDecoder.Cumulator() {
    @Override
    public ByteBuf cumulate(ByteBufAllocator alloc, ByteBuf cumulation,
    ByteBuf in) {
        ByteBuf buffer;
        try {
            /**
             * 引用大于1，说明用户使用了 slice().retain()或 duplicate().retain()
             * 使 refCnt 增加且大于1
             * 此时扩容返回一个新的累积区 ByteBuf
             * 以便对老的累积区 ByteBuf 进行后续的处理
             */
            if (cumulation.refCnt() > 1) {
                buffer = expandCumulation(alloc, cumulation,
                                                in.readableBytes());
                buffer.writeBytes(in);
            } else {
                CompositeByteBuf composite;
                if (cumulation instanceof CompositeByteBuf) {
                    composite = (CompositeByteBuf) cumulation;
                } else {
                    //创建 CompositeByteBuf，把 cumulation 包装成 Component 元素
                    //并将其加入复合缓冲区中
                    composite = alloc.compositeBuffer(Integer.MAX_VALUE);
                    composite.addComponent(true, cumulation);
```

```
            }
            //把 ByteBuf 也加入复合缓冲区中
            composite.addComponent(true, in);
            //赋空值,由于 ByteBuf 加入复合缓冲区后没有调用 retain()方法,因此无须释放
            in = null;
            buffer = composite;
        }
        return buffer;
    } finally {
        /**
         * in 不为 null,说明调用了 buffer 的 writeBytes()方法
         * 此时必须释放内存,以防止内存泄漏
         */
        if (in != null) {
            in.release();
        }
    }
}
```

最后,CompositeByteBuf 还有一个重要的方法,即移除已读字节——discardReadComponents(),其方法解读如下:

```
/**
 * 移除缓冲区中的已读字节
 * @return
 */
public CompositeByteBuf discardReadComponents() {
    ensureAccessible();
    final int readerIndex = readerIndex();
    if (readerIndex == 0) {
        return this;
    }
    int writerIndex = writerIndex();
    //若读/写索引等于容量,则说明容量已使用完,全部释放即可
    if (readerIndex == writerIndex && writerIndex == capacity()) {
        for (int i = 0, size = componentCount; i < size; i++) {
            components[i].free();
```

```
        }
        lastAccessed = null;
        clearComps();
        setIndex(0, 0);
        adjustMarkers(readerIndex);
        return this;
    }
    int firstComponentId = 0;
    CompositeByteBuf.Component c = null;
    //从数组第一个元素开始遍历
    for (int size = componentCount; firstComponentId < size; firstComponentId++) {
        c = components[firstComponentId];
        /**
         * 若结束位置大于读索引,则说明还有 buf 未读,无须再继续处理
         * 否则需要释放
         */
        if (c.endOffset > readerIndex) {
            break;
        }
        c.free();
    }
    //一个都没释放
    if (firstComponentId == 0) {
        return this;
    }
    //最后一次访问时的子 Buffer,若元素都被释放了,则置空 la
    CompositeByteBuf.Component la = lastAccessed;
    if (la != null && la.endOffset <= readerIndex) {
        lastAccessed = null;
    }
    //从数组中移除已释放的元素
    removeCompRange(0, firstComponentId);
    // 更新读/写索引
    int offset = c.offset;
    //从第一个元素开始更新元素的位置信息
    updateComponentOffsets(0);
```

```
    setIndex(readerIndex - offset, writerIndex - offset);
    //更新标记索引
    adjustMarkers(offset);
    return this;
}
```

4.4.5　PooledByteBuf 源码剖析

下面介绍一个非常重要的 ByteBuf 抽象类——PooledByteBuf。这个类继承于 AbstractReference CountedByteBuf，其对象主要由内存池分配器 PooledByteBufAllocator 创建。比较常用的实现类有两种：一种是基于堆外直接内存池构建的 PooledDirectByteBuf，是 Netty 在进行 I/O 的读/写时的内存分配的默认方式，堆外直接内存可以减少内存数据拷贝的次数；另一种是基于堆内内存池构建的 PooledHeapByteBuf。

除了上述两种实现类，Netty 还使用 Java 的后门类 sun.misc.Unsafe 实现了两个缓冲区，即 PooledUnsafeDirectByteBuf 和 PooledUnsafeHeapByteBuf。这个强大的后门类会暴露对象的底层地址，一般不建议使用，Netty 为了优化性能引入了 Unsafe。

PooledUnsafeDirectByteBuf 会暴露底层 DirectByteBuffer 的地址 memoryAddress，而 Unsafe 则可通过 memoryAddress+Index 的方式取得对应的字节。

PooledUnsafeHeapByteBuf 也会暴露字节数组在 Java 堆中的地址，因此不再使用字节数组的索引，即 array[index]。同样，Unsafe 可通过 BYTE_ARRAY_BASE_OFFSET+Index 字节的地址获取字节。

由于创建 PooledByteBuf 对象的开销大，而且在高并发情况下，当网络 I/O 进行读/写时会创建大量的实例。因此，为了降低系统开销，Netty 对 Buffer 对象进行了池化，缓存了 Buffer 对象，使对此类型的 Buffer 可进行重复利用。PooledByteBuf 是从内存池中分配出来的 Buffer，因此它需要包含内存池的相关信息，如内存块 Chunk、PooledByteBuf 在内存块中的位置及其本身所占空间的大小等。图 4-11 描述了 PooledByteBuf 的核心功能和属性。接下来对这些功能的源码进行详细的解读。

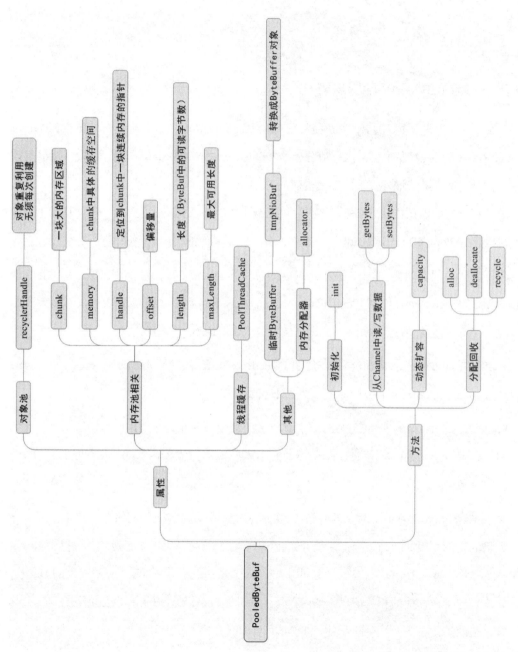

图 4-11 PooledByteBuf 的核心功能和属性

PooledByteBuf 的初始化，以及从 Channel 中读/写数据的解读如下：

```
private void init0(PoolChunk<T> chunk, ByteBuffer nioBuffer,
                   long handle, int offset, int length, int maxLength,
 PoolThreadCache cache) {
        assert handle >= 0;
        assert chunk != null;
        //大内存块默认为 16MB，被分配给多个 PooledByteBuf
        this.chunk = chunk;
        //chunk 中具体的缓存空间
        memory = chunk.memory;
        //将 PooledByteBuf 转换成 ByteBuffer
        tmpNioBuf = nioBuffer;
        //内存分配器：PooledByteBuf 是由 Arena 的分配器构建的
        allocator = chunk.arena.parent;
        //线程缓存：优先从线程缓存中获取
        this.cache = cache;
        //通过这个指针可以得到 PooledByteBuf 在 chunk 这棵二叉树中的具体位置
        this.handle = handle;
        //偏移量
        this.offset = offset;
        //长度：实际数据长度
        this.length = length;
        //写指针不能超过 PooledByteBuf 的最大可用长度
        this.maxLength = maxLength;
}
    /**
     * channel 从 PooledByteBuf 中获取数据
     * PooledByteBuf 的读索引的变化
     * 由父类 AbstractByteBuf 的 readBytes()方法维护
     */
   public final int getBytes(int index, GatheringByteChannel out, int length)
 throws IOException {
        return out.write(duplicateInternalNioBuffer(index, length));
   }
    /**
```

```java
 * 从 memory 中创建一份缓存 ByteBuffer
 * 与 memory 共享底层数据，但读/写索引独立维护
 */
ByteBuffer duplicateInternalNioBuffer(int index, int length) {
    //检查
    checkIndex(index, length);
    return _internalNioBuffer(index, length, true);
}
final ByteBuffer _internalNioBuffer(int index, int length,
boolean duplicate) {
    //获取读索引
    index = idx(index);
    //当 duplicate 为 true 时，在 memory 中创建共享此缓冲区内容的新的字节缓冲区
    //当 duplicate 为 false 时，先从 tmpNioBuf 中获取，当 tmpNioBuf 为空时
    //再调用 newInternalNioBuffer，此处与 memory 的类型有关，因此其具体实现由子类完成
    ByteBuffer buffer = duplicate ? newInternalNioBuffer(memory)
                        : internalNioBuffer();
    //设置新的缓冲区指针位置及 limit
    buffer.limit(index + length).position(index);
    return buffer;
}
protected final ByteBuffer internalNioBuffer() {
    ByteBuffer tmpNioBuf = this.tmpNioBuf;
    if (tmpNioBuf == null) {
        this.tmpNioBuf = tmpNioBuf = newInternalNioBuffer(memory);
    }
    return tmpNioBuf;
}
public final ByteBuffer internalNioBuffer(int index, int length) {
    checkIndex(index, length);
    //只有当 tmpNioBuf 为空时才创建新的共享缓冲区
    return _internalNioBuffer(index, length, false);
}
/**
 * 从 Channel 中读取数据并写入 PooledByteBuf 中
 * writerIndex 由父类 AbstractByteBuf 的 writeBytes()方法维护
```

```
 */
public final int setBytes(int index, ScatteringByteChannel in,
                 int length) throws IOException {
    try {
        return in.read(internalNioBuffer(index, length));
    } catch (ClosedChannelException ignored) {
        //客户端主动关闭连接，返回-1，触发对应的用户事件
        return -1;
    }
}
```

自动扩容代码解读如下：

```
/**
 * 自动扩容
 * @param newCapacity 新的容量值
 * @return
 */
public final ByteBuf capacity(int newCapacity) {
    //若新的容量值与长度相等，则无须扩容，直接返回即可
    if (newCapacity == length) {
        ensureAccessible();
        return this;
    }
    //检查新的容量值是否大于最大允许容量
    checkNewCapacity(newCapacity);
    /**
     * 非内存池，在新容量值小于最大长度值的情况下，无须重新分配
     * 只需修改索引和数据长度即可
     */
    if (!chunk.unpooled) {
        /**
         * 新的容量值大于长度值
         * 在没有超过Buffer的最大可用长度值时，只需把长度设为新的容量值即可
         * 若超过了最大可用长度值，则只能重新分配
         */
        if (newCapacity > length) {
```

```
            if (newCapacity <= maxLength) {
                length = newCapacity;
                return this;
            }
        } else if (newCapacity > maxLength >>> 1 &&
            (maxLength > 512 || newCapacity > maxLength - 16)) {
            //当新容量值小于最大可用长度值时,其读/写索引不能超过新容量值
            length = newCapacity;
            setIndex(Math.min(readerIndex(), newCapacity),
                    Math.min(writerIndex(), newCapacity));
            return this;
        }
    }
    //由Arena重新分配内存并释放旧的内存空间
    chunk.arena.reallocate(this, newCapacity, true);
    return this;
}
```

PooledByteBuf 对象回收代码解读如下:

```
/**
 * 对象回收,把对象属性清空
 */
protected final void deallocate() {
    if (handle >= 0) {
        final long handle = this.handle;
        this.handle = -1;
        memory = null;
        //释放内存
        chunk.arena.free(chunk, tmpNioBuf, handle, maxLength, cache);
        tmpNioBuf = null;
        chunk = null;
        recycle();
    }
}
/**
 * 把 PooledByteBuf 放回对象池 Stack 中,以便下次使用
```

```
    */
    private void recycle() {
        recyclerHandle.recycle(this);
    }
```

本小节只对 PooledByteBuf 的内部属性和部分方法进行了解读，至于 PoolChunk、PoolArena 及其内存池分配和回收的底层实现原理的详细剖析，会在第 6 章进行讲解。

4.5　Netty 内存泄漏检测机制源码剖析

Netty 在默认情况下采用的是池化的 PooledByteBuf，以提高程序性能。但是 PooledByteBuf 在使用完毕后需要手动释放，否则会因 PooledByteBuf 申请的内存空间没有归还导致内存泄漏，最终使内存溢出。一旦泄漏发生，在复杂的应用程序中找到未释放的 ByteBuf 并不是一个简单的事，在没有工具辅助的情况下只能检查所有源码，效率很低。

为了解决这个问题，Netty 运用 JDK 的弱引用和引用队列设计了一套专门的内存泄漏检测机制，用于实现对需要手动释放的 ByteBuf 对象的监控。

图 4-12 描述了各种引用及其各自的特征。

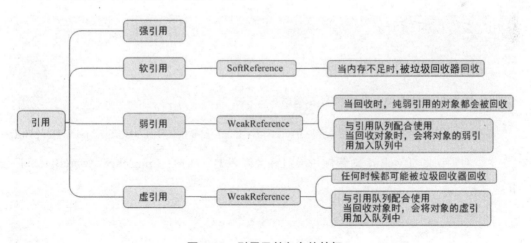

图 4-12　引用及其各自的特征

强引用：经常使用的编码方式，如果将一个对象赋值给一个变量，只要这个变量可用，那么这个对象的值就被该变量强引用了；否则垃圾回收器不会回收该对象。

软引用：当内存不足时，被垃圾回收器回收，一般用于缓存。

弱引用：只要是发生回收的时候，纯弱引用的对象都会被回收；当对象未被回收时，弱引用可以获取引用的对象。

虚引用：在任何时候都可能被垃圾回收器回收。如果一个对象与虚引用关联，则该对象与没有引用与之关联时一样。虚引用获取不到引用的对象。

引用队列：与虚引用或弱引用配合使用，当普通对象被垃圾回收器回收时，会将对象的弱引用和虚引用加入引用队列中。Netty 运用这一特性来检测这些被回收的 ByteBuf 是否已经释放了内存空间。下面对其实现原理及源码进行详细剖析。

4.5.1 内存泄漏检测原理

Netty 的内存泄漏检测机制主要是检测 ByteBuf 的内存是否正常释放。想要实现这种机制，就需要完成以下 3 步。

（1）采集 ByteBuf 对象。

（2）记录 ByteBuf 的最新调用轨迹信息，方便溯源。

（3）检查是否有泄漏，并进行日志输出。

第一，采集入口在内存分配器 PooledByteBufAllocator 的 newDirectBuffer 与 newHeapBuffer 方法中，对返回的 ByteBuf 对象做一层包装，包装类分两种：SimpleLeakAwareByteBuf 与 AdvancedLeakAwareByteBuf。

AdvancedLeakAwareByteBuf 是 SimpleLeakAwareByteBuf 的子类，它们的主要作用都是记录 ByteBuf 的调用轨迹。区别在于，AdvancedLeakAwareByteBuf 记录 ByteBuf 的所有操作；SimpleLeakAwareByteBuf 只在 ByteBuf 被销毁时告诉内存泄漏检测工具把正常销毁的对象从检测缓存中移除，方便判断 ByteBuf 是否泄漏，不记录 ByteBuf 的操作。

第二，每个 ByteBuf 的最新调用栈信息记录在其弱引用中，这个弱引用对象与 ByteBuf 都包装在 SimpleLeakAwareByteBuf 类中。弱引用对象除了记录 ByteBuf 的调用轨迹，还要有关闭检测的功能，因为当 ByteBuf 被销毁时需要关闭资源跟踪，并清除对资源对象的引用，防止误报。

第三，在创建弱引用时，需要引用队列的配合。当检测是否有资源泄漏时，需要遍历引用队列，找到已回收的 ByteBuf 的引用，通过这些引用判断是否调用了 ByteBuf 的销毁接口，检测是否有泄漏。

Netty 先把所有弱引用缓存起来，在 ByteBuf 被销毁后，再从缓存中移除对应的弱引用，当遍历到此弱引用时，若发现它已从缓存中移除，则表示 ByteBuf 无内存泄漏。此种判断方式有点特别，一般只需在类中加个原子属性即可。例如，AtomicBoolean，在将其正常销毁后，把弱引用的原子属性值设为 true，当检测判断时，若此原子属性为 false，则表示非正常销毁。但 Netty 未采用这种方式，而是使用缓存容器来判断是否有泄漏。Netty 会把这些弱引用对象强引用起来。由于 ByteBuf 资源对象被垃圾回收器回收后，若其弱引用对象若没有地方强关联，则会在下一次被垃圾回收器回收，因此 Netty 采用全局 Set 把它缓存起来，防止弱引用对象在遍历之前被回收。

4.5.2　内存泄漏器 ResourceLeakDetector 源码剖析

ResourceLeakDetector 在整个内存泄漏检测机制中起核心作用。一种缓冲区资源会创建一个 ResourceLeakDetector 实例，并监控此缓冲区类型的池化资源（本书只介绍 AbstractByteBuf 类型的资源）。ResourceLeakDetector 的 trace() 方法是整套检测机制的入口，提供资源采集逻辑，运用全局的引用队列和引用缓存 Set 构建 ByteBuf 的弱引用对象，并检测当前监控的资源是否出现了内存泄漏。若出现了内存泄漏，则输出泄漏报告及内存调用轨迹信息。

ResourceLeakDetector 中有个私有类——DefaultResourceLeak，实现了 ResourceLeakTracker 接口，主要负责跟踪资源的最近调用轨迹，同时继承 WeakReference 弱引用。调用轨迹的记录被加入 DefaultResourceLeak 的 Record 链表中，Record 链表不会保存所有记录，因为它的长度有一定的限制。

Netty 的内存泄漏检测机制有以下 4 种检测级别。

- DISABLED：表示禁用，不开启检测。
- SIMPLE：Netty 的默认设置，表示按一定比例采集。若采集的 ByteBuf 出现泄漏，则打印 LEAK:XXX 等日志，但没有 ByteBuf 的任何调用栈信息输出，因为它使用的包装类是 SimpleLeakAwareByteBuf，不会进行记录。
- ADVANCED：它的采集与 SIMPLE 级别的采集一样，但会输出 ByteBuf 的调用栈信息，因为它使用的包装类是 AdvancedLeakAwareByteBuf。
- PARANOID：偏执级别，这种级别在 ADVANCED 的基础上按 100%的比例采集。

当系统处于开发和功能测试阶段时，一般会把级别设置为 PARANOID，容易发现问题。在系统正式上线后，会把级别降到 SIMPLE。若出现了泄漏日志的情况，则在重启服务时，可以把级别调为 ADVANCED，查找内存泄漏的轨迹，方便定位。当系统上线很长一段时间后，比较稳定了，可以禁用内存泄漏检测机制。Netty 对这些级别的处理具体是怎样实现按一定比例采集的呢？通过接下来的源码解读来寻找答案。图 4-13 为内存泄漏检测机制的功能。

第 4 章 Netty 核心组件源码剖析

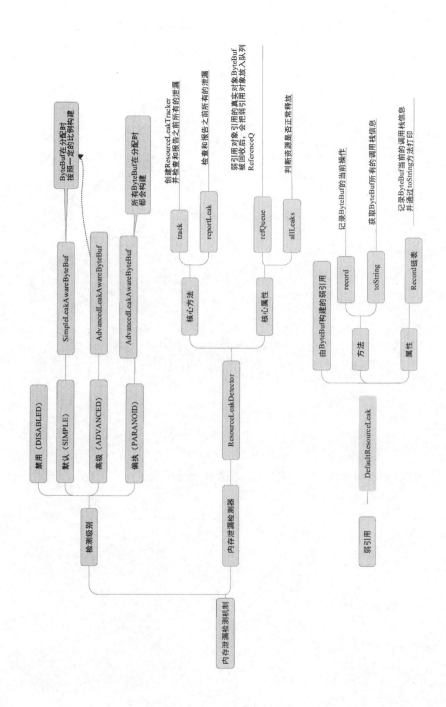

图 4-13 内存泄漏检测机制的功能

源码解读先从入口 AbstractByteBufAllocator 的 toLeakAwareBuffer()方法开始，具体代码解读如下：

```java
/**
 * 内存泄漏检测入口
 * ByteBuf 在分配后需要交给内存泄漏检测器处理
 * 处理完后对 ByteBuf 对象进行相应的包装并返回
 * @param buf
 * @return
 */
protected static ByteBuf toLeakAwareBuffer(ByteBuf buf) {
    //弱引用
    ResourceLeakTracker<ByteBuf> leak;
    switch (ResourceLeakDetector.getLevel()) {
        //默认级别
        case SIMPLE:
            /**
             * 每种类型的资源都会创建一个内存泄漏检测器 ResourceLeakDetector
             * 通过内存泄漏检测器获取弱引用
             */
            leak = AbstractByteBuf.leakDetector.track(buf);
            /**
             * 若弱引用不为空，则说明此 buf 被采集了
             * buf 一旦被采集
             * 就需要返回对应级别的包装对象，否则会出现误报
             */
            if (leak != null) {
                buf = new SimpleLeakAwareByteBuf(buf, leak);
            }
            break;
        /**
         * 高级和偏执级别
         * 由于它们都需要追踪 buf 的调用轨迹
         * 因此返回的包装对象相同
         */
        case ADVANCED:
        case PARANOID:
```

```
            leak = AbstractByteBuf.leakDetector.track(buf);
            if (leak != null) {
                buf = new AdvancedLeakAwareByteBuf(buf, leak);
            }
            break;
        default:
            break;
    }
    return buf;
}
```

重点关注内存泄漏检测器的 track() 方法，此方法不仅采集 buf，还会在采集完后，检测是否有内存泄漏的 buf，并打印日志。具体代码解读如下：

```
public final ResourceLeakTracker<T> track(T obj) {
    return track0(obj);
}
private ResourceLeakDetector.DefaultResourceLeak track0(T obj) {
    //获取内存泄漏检测级别
    ResourceLeakDetector.Level level = ResourceLeakDetector.level;
    //不检测，也不采集
    if (level == ResourceLeakDetector.Level.DISABLED) {
        return null;
    }
    /**
     * 当级别比偏执级别低时
     * 获取一个 128 以内的随机数
     * 若得到的数不为 0，则不采集
     * 若为 0，则检测是否有泄漏，并输出泄漏日志
     * 同时创建一个弱引用
     */
    if (level.ordinal() < ResourceLeakDetector.Level.PARANOID.ordinal()) {
        if((PlatformDependent.threadLocalRandom().
                nextInt(samplingInterval)) == 0) {
            reportLeak();
          return new ResourceLeakDetector.DefaultResourceLeak(obj,
                                            refQueue, allLeaks);
        }
```

```
            return null;
    }
    //偏执级别都采集
    reportLeak();
    return new ResourceLeakDetector.DefaultResourceLeak(obj,refQueue, allLeaks);
}
private void reportLeak() {
    if (!logger.isErrorEnabled()) {
        clearRefQueue();
        return;
    }
    /**
     * 循环获取引用队列中的弱引用
     */
    for (;;) {
        ResourceLeakDetector.DefaultResourceLeak ref =
                (ResourceLeakDetector.DefaultResourceLeak)
                                                refQueue.poll();
        if (ref == null) {
            break;
        }
        //检测是否泄漏
        //若未泄漏，则继续下一次循环
        if (!ref.dispose()) {
            continue;
        }
        //获取 buf 的调用栈信息
        String records = ref.toString();
        //不再输出曾经输出过的泄漏记录
        if (reportedLeaks.putIfAbsent(records, Boolean.TRUE) == null) {
            if (records.isEmpty()) {
                reportUntracedLeak(resourceType);
            } else {
                //输出内存泄漏日志及其调用栈信息
                reportTracedLeak(resourceType, records);
            }
        }
```

```
        }
    }
    /**
     * 判断是否泄漏
     */
    boolean dispose() {
        //清理对资源对象的引用
        clear();
        //若引用缓存中还存在此引用,则说明 buf 未释放,内存泄漏了
        return allLeaks.remove(this);
    }
    protected void reportTracedLeak(String resourceType, String records) {
        logger.error(
                "LEAK: {}.release() was not called before it's garbage-collected. " +"See https://netty.io/wiki/reference-counted-objects.html for more information.{}",resourceType, records);
    }
    /**
     * 弱引用重写了 toString()方法
     * 需注意:若采用 IDE 工具 debug 调试代码
     * 则在处理对象时,IDE 会自动调用 toString()方法
     * @return
     */
    @Override
    public String toString() {
        //获取记录列表的头部
        ResourceLeakDetector.Record oldHead = headUpdater.getAndSet(this,null);
        //若无记录,则返回空字符串
        if (oldHead == null) {
            return EMPTY_STRING;
        }
        //若记录太长,则会丢弃部分记录,获取丢弃了多少记录
        final int dropped = droppedRecordsUpdater.get(this);
        int duped = 0;
        /**
         * 由于每次在链表新增头部时,其 pos=旧的 pos+1
         * 因此最新的链表头部的 pos 就是链表长度
```

```java
 */
int present = oldHead.pos + 1;
//设置 buf 的容量（大概为 2KB 栈信息*链表长度），并添加换行符
StringBuilder buf = new StringBuilder(present * 2048).append(NEWLINE);
buf.append("Recent access records: ").append(NEWLINE);
int i = 1;
Set<String> seen = new HashSet<String>(present);
for (; oldHead != ResourceLeakDetector.Record.BOTTOM; oldHead =
 oldHead.next) {
    //获取调用栈信息
    String s = oldHead.toString();
    if (seen.add(s)) {
        /**
         * 遍历到最初的记录与其他节点的输出有所不同
         */
        if (oldHead.next == ResourceLeakDetector.Record.BOTTOM) {
            buf.append("Created at:").append(NEWLINE).append(s);
        } else {
    buf.append('#').append(i++).append(':').append(NEWLINE).append(s);
        }
    } else {
        //出现重复的记录
        duped++;
    }
}
/**
 * 当出现重复的记录时，加上特殊日志
 */
if (duped > 0) {
    buf.append(": ")
        .append(duped)
        .append(" leak records were discarded because they were
         duplicates")
        .append(NEWLINE);
}
/**
 * 若出现记录数超过了 TARGET_RECORDS（默认为 4）
```

```
   * 则输出丢弃了多少记录等额外信息
   * 可通过设置 io.netty.leakDetection.targetRecords 来修改记录长度
   */
  if (dropped > 0) {
     buf.append(": ")
           .append(dropped)
           .append(" leak records were discarded
                 because the leak record count is targeted to ")
           .append(TARGET_RECORDS)
           .append(". Use system property ")
           .append(PROP_TARGET_RECORDS)
           .append(" to increase the limit.")
           .append(NEWLINE);
  }
  buf.setLength(buf.length() - NEWLINE.length());
  return buf.toString();
}
```

在 ADVANCED 之上的级别的操作中,ByteBuf 的每项操作都涉及线程调用栈轨迹的记录。那么该如何获取线程栈调用信息呢？在记录某个点的调用栈信息时,Netty 会创建一个 Record 对象,Record 类继承 Exception 的父类 Throwable。因此在创建 Record 对象时,当前线程的调用栈信息就会被保存起来。关于调用栈信息的保存及获取的代码解读如下：

```
/**
 * 记录调用轨迹
 * @param hint 提示对象
 */
private void record0(Object hint) {
    //如果 TARGET_RECORDS 大于 0, 则记录
    if (TARGET_RECORDS > 0) {
        ResourceLeakDetector.Record oldHead;
        ResourceLeakDetector.Record prevHead;
        ResourceLeakDetector.Record newHead;
        boolean dropped;
        do {
            /**
             * 判断记录链头是否为空,为空表示已关闭
```

```
 * 把之前的链头作为第二个元素赋值给新链表
 */
if ((prevHead = oldHead = headUpdater.get(this)) == null) {
    return;
}
//获取链表长度
final int numElements = oldHead.pos + 1;
//若链表长度大于或等于最大长度值 TARGET_RECORDS
if (numElements >= TARGET_RECORDS) {
    /**
     * backOffFactor 是用来计算是否替换的因子
     * 其最小值为 numElements-TARGET_RECORDS
     * 元素越多，其值越大，最大值为 30
     */
    final int backOffFactor = Math.min(numElements - TARGET_RECORDS, 30);
    /**
     * 1/2^backOffFactor 的概率不会执行此 if 代码块
     * 代码块：prevHead=oldHead.next
     * 表示用之前链头元素作为新链表的第二个元素
     * 丢弃原来的链头
     * 同时设置 dropped 为 false
     */
    if (dropped = PlatformDependent.threadLocalRandom().nextInt(1
            << backOffFactor) != 0) {
        prevHead = oldHead.next;
    }
} else {
    dropped = false;
}
//创建一个新的 Record，并将其添加到链表上，作为链表新的头部
newHead = hint != null ? new ResourceLeakDetector.Record(prevHead,
            hint): new ResourceLeakDetector.Record(prevHead);
} while (!headUpdater.compareAndSet(this, oldHead, newHead));
//若有丢弃，则更新记录丢弃的数
if (dropped) {
    droppedRecordsUpdater.incrementAndGet(this);
}
```

```java
    }
}
/**
 * Record 的 toString()方法
 * 获取 Record 创建时的调用栈信息
 * @return
 */
public String toString() {
    StringBuilder buf = new StringBuilder(2048);
    //先添加提示信息
    if (hintString != null) {
        buf.append("\tHint: ").append(hintString).append(NEWLINE);
    }
    //再添加栈信息
    StackTraceElement[] array = getStackTrace();
    // 跳过前面 3 个栈元素
    // 因为它们是 record()方法的栈信息,显示没意义
    out: for (int i = 3; i < array.length; i++) {
        StackTraceElement element = array[i];
        //跳过一些不必要的方法信息
        String[] exclusions = excludedMethods.get();
        for (int k = 0; k < exclusions.length; k += 2) {
            if (exclusions[k].equals(element.getClassName())
                    && exclusions[k + 1].equals(element.getMethodName())) {
                continue out;
            }
        }
        //格式化
        buf.append('\t');
        buf.append(element.toString());
        //加上换行
        buf.append(NEWLINE);
    }
    return buf.toString();
}
```

至此,整个内存泄漏检测机制的源码已经解读完了,通过学习 Netty 的这套内存泄漏检测

机制，不仅了解了如何检测内存是否泄漏，还对栈信息的调用保存和弱引用、虚引用的应用有了一定的了解。在研究源码时，建议对不太理解的代码编写 Junit 测试进行调试。下面是弱引用的测试用例：

```
@Test
public void testWeekReference() throws InterruptedException {
    // 创建字符串对象
    String str = new String("内存泄漏检测");
    // 创建一个引用队列
    ReferenceQueue referenceQueue = new ReferenceQueue();
    // 创建一个弱引用，弱引用引用 str 字符串
    WeakReference weakReference = new WeakReference (str , referenceQueue);
    // 切断 str 引用和"内存泄漏检测"字符串之间的引用
    str = null;
    // 取出弱引用所引用的对象
    // 若是虚引用，则虚引用无法获取被引用的对象
    System.out.println(weakReference.get());
    System.gc();
    // 强制垃圾回收
    System.runFinalization();
    // 垃圾回收之后，弱引用将被放入引用队列
    // 取出引用队列中的引用并与 weakReference 进行比较，应输出 true
    System.out.println(referenceQueue.poll() == weakReference);
}
```

4.6 小结

本章主要对 Netty 的 NioEventLoop 线程、Channel、ByteBuf 缓冲区、内存泄漏检测机制进行了详细的剖析。这些组件都是 Netty 的核心，在学习其源码的同时，要思考其整体设计思想，如 Channel 和 ByteBuf 的整体设计及其每层抽象类的意义；在这些组件中，Netty 对哪些部分做了性能优化，如运用 JDK 的 Unsafe、乐观锁、对象池、SelectedSelectionKeySet 数据结构优化等。除了本章的核心组件，Netty 还有 Handler 事件驱动模型、编码和解码、时间轮、复杂的内存池管理等，在后续章节会进行详细的剖析。

Netty 读/写请求源码剖析

5.1 ServerBootstrap 启动过程剖析

本节主要结合实际应用，联合第 4 章分析的核心组件，对 Netty 的整体运行机制进行详细的剖析，主要分为两部分：第一部分，Netty 服务的启动过程及其内部线程处理接入 Socket 链路的过程；第二部分，Socket 链路数据的读/写。本节主要通过普通 Java NIO 服务器的设计流程来详细讲述 Netty 是如何运用 NIO 启动服务器，并完成端口监听以接收客户端的请求的；以及它与直接使用 NIO 有哪些区别，有什么优点。下面先看看 NIO 中几个比较重要的类。

- Selector：多路复用器，是 NIO 的一个核心组件，在 BIO 中是没有的，主要用于监听 NIO Channel（通道）的各种事件，并检查一个或多个通道是否处于可读/写状态。在实现单条线程管理多条链路时，传统的 BIO 编程管理多条链路都是通过多线程上下切换来实现的，而 NIO 有了 Selector 后，一个 Selector 只使用一条线程就可以轮询处理多条链路。

- ServerSocketChannel：与普通 BIO 中的 ServerSocket 一样，主要用来监听新加入的 TCP 连接的通道，而且其启动方式与 ServerSocket 的启动方式也非常相似，只需要在开启端口监听之前，把 ServerSocketChannel 注册到 Selector 上，并设置监听 OP_ACCEPT 事件即可。
- SocketChannel：与普通 BIO 中的 Socket 一样，是连接到 TCP 网络 Socket 上的 Channel。它的创建方式有两种：一种方式与 ServerSocketChannel 的创建方式类似，打开一个 SocketChannel，这种方式主要用于客户端主动连接服务器；另一种方式是当有客户端连接到 ServerSocketChannel 上时，会创建一个 SocketChannel。

NIO 服务器设计思维导图如图 5-1 所示，主要是关于编写一套 NIO 服务器需要完成的一些基本步骤，主要分三步：第一步，服务器启动与端口监听；第二步，ServerSocketChannel 处理新接入链路；第三步，SocketChannel 读/写数据。

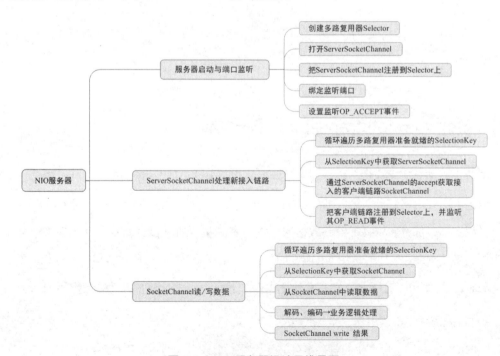

图 5-1　NIO 服务器设计思维导图

第 5 章　Netty 读/写请求源码剖析

　　Netty 服务的启动流程与图 5-1 中的 NIO 服务器启动的核心部分没什么区别，同样需要创建 Selector，不过它会开启额外线程去创建；同样需要打开 ServerSocketChannel，只是采用的是 NioServerSocketChannel 来进行包装；同样需要把 ServerSocketChannel 注册到 Selector 上，只是这些事情都是在额外的 NioEventLoop 线程上执行的，并返回 ChannelPromise 来异步通知是否注册成功。ChannelPromise 在之前的实战中运用过（Netty 在多线程操作中会大量使用）。Netty 服务启动时涉及的类和方法如图 5-2 所示，接下来通过几个问题，由浅入深地进行剖析。

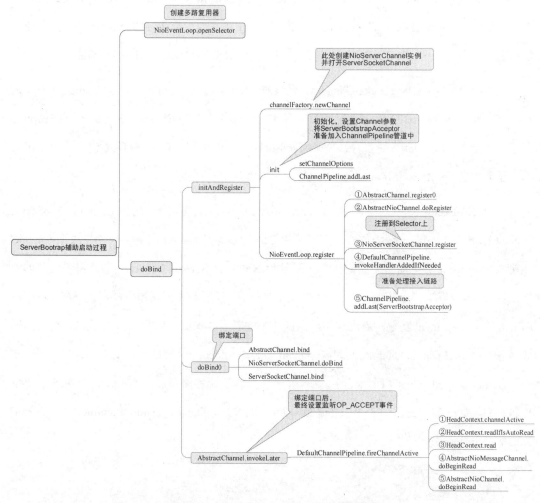

图 5-2　Netty 服务启动时涉及的类和方法

167

Netty 服务的启动主要分以下 5 步。

（1）创建两个线程组，并调用父类 MultithreadEventExecutorGroup 的构造方法实例化每个线程组的子线程数组，Boss 线程组只设置一条线程，Worker 线程组默认线程数为 Netty Runtime.availableProcessors() * 2。在 NioEventLoop 线程创建的同时多路复用器 Selector 被开启（每条 NioEventLoop 线程都会开启一个多路复用器）。

（2）在 AbstractBootstrap 的 initAndRegister 中，通过 ReflectiveChannelFactory.newChannel() 来反射创建 NioServerSocketChannel 对象。由于 Netty 不仅仅只提供 TCP NIO 服务，因此此处使用了反射开启 ServerSocketChannel 通道，并赋值给 SelectableChannel 的 ch 属性。

（3）初始化 NioServerSocketChannel、设置属性 attr 和参数 option，并把 Handler 预添加到 NioServerSocketChannel 的 Pipeline 管道中。其中，attr 是绑定在每个 Channel 上的上下文；option 一般用来设置一些 Channel 的参数；NioServerSocketChannel 上的 Handler 除了包括用户自定义的，还会加上 ServerBootstrapAcceptor。

（4）NioEventLoop 线程调用 AbstractUnsafe.register0() 方法，此方法执行 NioServerSocketChannel 的 doRegister() 方法。底层调用 ServerSocketChannel 的 register() 方法把 Channel 注册到 Selector 上，同时带上了附件，此附件为 NioServerSocketChannel 对象本身。此处的附件 attachment 与第（3）步的 attr 很相似，在后续多路复用器轮询到事件就绪的 SelectionKey 时，通过 k.attachment 获取。当出现超时或链路未中断或移除时，JVM 不会回收此附件。注册成功后，会调用 DefaultChannelPipeline 的 callHandlerAddedForAllHandlers() 方法，此方法会执行 PendingHandlerCallback 回调任务，回调原来在没有注册之前添加的 Handler。此处有点难以理解，在注册之前，先运行了 Pipeline 的 addLast() 方法。DefaultChannelPipeline 的 addLast() 方法的部分代码如下：

```
if (!this.registered) {
 newCtx.setAddPending();
 this.callHandlerCallbackLater(newCtx, true);
```

```
return this;
}
```

上述回调任务主要是把 ServerBootstrapAcceptor 和自定义的 Handler 加入 NioServerSocket 的 Pipeline 管道中。

（5）注册成功后会触发 ChannelFutureListener 的 operationComplete()方法，此方法会带上主线程的 ChannelPromise 参数，然后调用 AbstractChannel.bind()方法；再执行 NioServerSocketChannel 的 doBind()方法绑定端口；当绑定成功后，会触发 active 事件，为注册到 Selector 上的 ServerSocket Channel 加上监听 OP_ACCEPT 事件；最终运行 ChannelPromise 的 safeSetSuccess()方法唤醒 serverBootstrap.bind(port).sync()。

以上 5 步是对图 5-2 进行的详细说明，Netty 服务的启动过程看起来涉及的类非常多，而且很多地方涉及多线程的交互（有主线程，还有 EventLoop 线程）。但由于 NioServerSocketChannel 通道绑定了一条 NioEventLoop 线程，而这条 NioEventLoop 线程上开启了 Selector 多路复用器，因此这些主要步骤的具体完成工作都会交给 NioEventLoop 线程，主线程只需完成协调和初始化工作即可。主线程通过 ChannelPromise 获取 NioEventLoop 线程的执行结果。这里有两个问题需要额外思考。

- ServerSocketChannel 在注册到 Selector 上后为何要等到绑定端口才设置监听 OP_ACCEPT 事件？提示：跟 Netty 的事件触发模型有关。
- NioServerSocketChannel 的 Handler 管道 DefaultChannelPipeline 是如何添加 Handler 并触发各种事件的？

这两个问题与 Netty 的架构设计有很大的关系，对于初学者来说，有一定的难度，一定要跟着图 5-2 及以上 5 步多看几遍源码，多加思考。

AbstractBoostrap 与 ServerBootstrap 初始化 Channel 并注册到 NioEventLoop 线程上，以及端口绑定的核心源码解读如下：

```
private ChannelFuture doBind(final SocketAddress localAddress) {
    //初始化 Channel 并注册到 NioEventLoop 线程上
```

```java
final ChannelFuture regFuture = initAndRegister();
final Channel channel = regFuture.channel();
//判断是否存在注册异常
if (regFuture.cause() != null) {
    return regFuture;
}
if (regFuture.isDone()) {
    //注册成功后需要绑定端口
    //由 NioEventLoop 线程去异步执行,此时需要创建 ChannelPromise 对象
    ChannelPromise promise = channel.newPromise();
    //最终调用 AbstractChannel 的 bind()方法
    doBind0(regFuture, channel, localAddress, promise);
    return promise;
} else {
    /**
     * 由于注册操作由 NioEventLoop 线程去异步执行,因此可能会执行不完
     * 此时需要返回 PendingRegistrationPromise 对象,及时把结果交互给主线程
     */
    final AbstractBootstrap.PendingRegistrationPromise promise =
            new AbstractBootstrap.PendingRegistrationPromise(channel);
    //加上注册监听器,注册动作完成后触发
    regFuture.addListener(new ChannelFutureListener() {
        @Override
        public void operationComplete(ChannelFuture future)
        throws Exception {
            Throwable cause = future.cause();
            if (cause != null) {
                //注册失败处理,响应给主线程
                promise.setFailure(cause);
            } else {
                //只有注册成功后才能绑定
                promise.registered();
                doBind0(regFuture, channel, localAddress, promise);
            }
        }
    });
    return promise;
```

```java
    }
}
final ChannelFuture initAndRegister() {
    Channel channel = null;
    try {
        /**
         * 根据 serverBootstrap.channel(NioServerSocketChannel.class)
         * 反射创建 NioServerSocketChannel 对象
         */
        channel = channelFactory.newChannel();
        /**
         * 初始化 NioServerSocketChannel
         * 设置 Channel 的参数, 为 Worker 线程管理的 SocketChannel 准备好参数
         * 及其 Handler 消息处理链
         */
        init(channel);
    } catch (Throwable t) {
        //初始化处理失败
        // 创建 DefaultChannelPromise 实例,设置异常并返回
        if (channel != null) {
            channel.unsafe().closeForcibly();
            return new DefaultChannelPromise(channel,
                    GlobalEventExecutor.INSTANCE).setFailure(t);
        }
        return new DefaultChannelPromise(new FailedChannel(),
                GlobalEventExecutor.INSTANCE).setFailure(t);
    }
    /**
     * 调用 SingleThreadEventLoop 的 register()方法
     * 最终触发了 AbstractUnsafe 的 register
     */
    ChannelFuture regFuture = config().group().register(channel);
    //注册异常处理
    if (regFuture.cause() != null) {
        if (channel.isRegistered()) {
            channel.close();
        } else {
```

```
            channel.unsafe().closeForcibly();
        }
    }
    /**
     * 这段注释比较长，主要讲述 Channel 注册成功后的一些操作
     * bind 或 connect 操作需要在 register 完成后执行
     * 此处涉及线程切换，因为 ServerBootStrap 运行在主线程上
     * 而 register、bind、connect 需要在 NioEventLoop 线程上执行
     * 注释翻译如下：
     * 如果程序到这里，则说明 promise 没有失败，可能发生以下情况
     * （1） 如果尝试将 Channel 注册到 EventLoop 上，且此时注册已经完成
     *       则 inEventLoop 返回 true，Channel 已经注册成功
     *       可以安全调用 bind() 或 connect()
     * （2） 如果尝试注册到另一个线程上，即 inEventLoop 返回 false
     *       则此时 register 请求已成功添加到事件循环的任务队列中
     *       现在同样可以尝试调用 bind() 或 connect()
     * 因为 register()、bind() 和 connect() 都被绑定在同一个 I/O 线程上
     * 所以在执行完 register Task 之后，bind() 或 connect() 才会被执行
     */
    return regFuture;
}
```

5.2　Netty 对 I/O 就绪事件的处理

5.2.1　NioEventLoop 就绪处理之 OP_ACCEPT

通过前面的学习，了解了 Netty 服务的启动过程，以及 Netty 服务采用辅助类 ServerBootstrap 启动 NioEventLoop 线程，并依次开启 Selector、创建 ServerSocketChannel 并注册到 Selector 上、设置监听 OP_ACCEPT 事件的过程。那么当有 Socket 通道接入时，Netty 是如何处理的呢？本节还是通过图、文字及 Netty 部分源码的方式对这块处理逻辑进行详细的剖析。下面先看一幅 NioEventLoop 处理就绪 OP_ACCEPT 事件的时序图，如图 5-3 所示。

第 5 章 Netty 读/写请求源码剖析

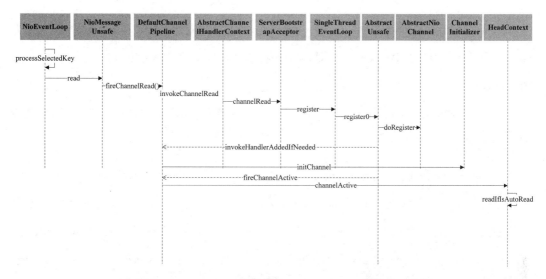

图 5-3 NioEventLoop 处理就绪 OP_ACCEPT 事件的时序图

图 5-3 主要分以下 3 步。

（1）当 NioEventLoop 中的多路复用器 Selector 轮询到就绪的 SelectionKey 时，判断 Key 的 readyOps 类型是否为 OP_ACCEPT，若是，则 5.1 节提到的 Key 的 attachment 就是 NioServerSocketChannel 本身，先获取 SelectionKey 的 attachment 对象，再触发此对象的辅助类 Unsafe 的实现类 NioMessageUnsafe 的 read()方法进行处理。

（2）在 NioMessageUnsafe 的 read()方法中会执行 doReadMessages（此处用到了模板设计模式）。真正调用的是 AbstractNioMessageChannel 的子类 NioServerSocketChannel 的 doReadMessages()方法。此方法最终调用 ServerSocketChannel 的 accept()方法，以获取接入的 SocketChannel。将 accept()方法在 AbstractNioChannel 的构造方法中设置为非阻塞状态，不管是否有 Channel 接入，都会立刻返回，并且一次最多默认获取 16 个，可以通过设置 option 参数 MAX_MESSAGES_PER_READ 来调整。获取到 SocketChannel 后，构建 NioSocketChannel，并把构建好的 NioSocketChannel 对象作为消息 msg 传送给 Handler（此 Handler 是 ServerBootstrapAcceptor），触发 Pipeline 管道的 fireChannelRead()方法，进而触发 read 事件，最后会调用 Handler 的 channelRead()方法。

（3）在 ServerBootstrapAcceptor 的 channelRead()方法中，把 NioSocketChannel 注册到 Worker 线程上，同时绑定 Channel 的 Handler 链。这与 5.1 节中将 NioServerSocketChannel 注册到 Boss 线程上类似，代码流程基本上都一样，只是实现的子类不一样，如后续添加的事件由 OP_ACCEPT 换成了 OP_READ。通过这一步的分析，读者可以思考，Netty 为何要把 Channel 抽象化？

当将 NioSocketChannel 注册到 Selector 上时，有部分代码需要解读，NioSocketChannel 对应的 NioEventLoop 线程在未启动时，eventLoop.inEventLoop()会返回 false。若 Worker 的线程数为 16，则在前面 16 个 NioSocketChannel 注册时，都会把注册看作一个 Task 并添加到 NioEventLoop 的队列中，同时启动 NioEventLoop 队列，唤醒 Selector。这部分功能在 AbstractUnsafe 的 register()方法中，具体代码如下：

```
AbstractChannel.this.eventLoop = eventLoop;
if (eventLoop.inEventLoop()) {
register0(promise);
} else {
  try {
   eventLoop.execute(new OneTimeTask() {
      @Override
      public void run() {
         register0(promise);
      }
   });
}
```

在 AbstractUnsafe 的 register0()方法中有关于如何将用户自定义的 Hanlder 添加到 NioSocketChannel 的 Handler 链表中的方法，核心代码解读如下：

```
//此方法调用 AbstractNioChannel 的 doRegister()方法
把 NioServerSocketChannel 和 NioSocketChannel 的注册抽象出来
doRegister();
neverRegistered = false;
registered = true;
```

```
/**
 *在 ServerBootstrapAcceptor 的 channelRead()方法中把用户定义的 Handler 追加到 Chanel 的
 *管道中(child.pipeline().addLast(childHandler)),此方法会追加一个回调,此时正好会触发
 *这个回调
   if (!registered) {
   newCtx.setAddPending();
   callHandlerCallbackLater(newCtx, true);
   return this;
   }
 */
pipeline.invokeHandlerAddedIfNeeded();
//此方法会触发 promise 的监听
safeSetSuccess(promise);
pipeline.fireChannelRegistered();
if (isActive()) {
   if (firstRegistration) {
   //此方法会触发 HeadContext 的 channelActive()方法,并最终调用 AbstractNioChannel 的
   //doBeginRead()方法注册监听 OP_READ 事件
       pipeline.fireChannelActive();
   } else if (config().isAutoRead()) {
       beginRead();
   }
}
```

仔细阅读上述代码的注释,在 IDEA 上调试几遍。至此,NioEventLoop I/O 的读/写线程已开启,并一直轮询监听是否触发了 OP_READ 事件。下面会继续讲解 NioEventLoop 线程是如何读取客户端请求数据的。

5.2.2　NioEventLoop 就绪处理之 OP_READ(一)

通过前面的介绍,了解了 Boss 线程组中 NioEventLoop 线程处理 Socket 链路接入的整个过程。下面主要讲解 Worker 线程组 NioEventLoop 线程是如何读取 Socket 链路传过来的数据的。本小节跟 5.2.1 小节类似,区别在于实际处理 Unsafe 类从 NioMessageUnsafe 变成了 NioByteUnsafe,Handler 类变成了用户设置的编/解码器,以及业务逻辑处理 Handler 不再是 ServerBootstrapAcceptor。接下来先看一幅 NioEventLoop 处理就绪 OP_READ 事件的时

序图，如图 5-4 所示。

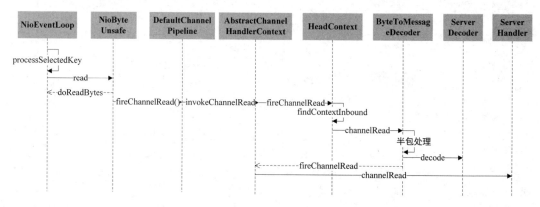

图 5-4　NioEventLoop 处理就绪 OP_READ 事件的时序图 1

图 5-4 与图 5-3 非常相似，在 5.2.1 小节中，NioMessageUnsafe 先调用 NioServerSocketChannel 的 doReadMessages()方法读取接入的 Channel。而本小节中的 NioByteUnsafe 不断地调用 NioSocketChannel 的 doReadBytes()方法从 Channel 中读取数据，再把读取到的 ByteBuf 交给管道 Pipeline，并触发后续一系列 ChannelInboundHandler 的 channelRead()方法。整个读取数据的过程涉及的 Handler 都是以 HeadContext 开头的，按顺序运行用户自定义的各个解码器和服务端业务逻辑处理 Handler。图 5-5 为 Netty 常用解码器类图。

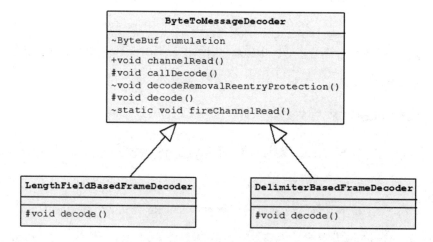

图 5-5　Netty 常用解码器类图

图 5-5 描述了 Netty 的解码器设计，采用了模板设计模式。对于用户选择的解码器，除了 MessageToMessageCodec 的子类，其他解码器首先都会经过其父类 ByteToMessageDecoder 的 channelRead()方法。channelRead()及相关方法的源码解读如下：

```
@Override
  public void channelRead(ChannelHandlerContext ctx, Object msg)
    throws Exception {
    if (msg instanceof ByteBuf) {
        //解码后的消息列表
        CodecOutputList out = CodecOutputList.newInstance();
        try {
            ByteBuf data = (ByteBuf) msg;
            //是否为第一次解码
            first = cumulation == null;
            if (first) {
                //在第一次解码时只需把 data（data 是 msg 的类型强转）赋给字节容器即可
                cumulation = data;
            } else {
                //若不是第一次解码，则需要把 msg 写入 cumulation 中
                //写入前需判断是否需要扩容
                cumulation = cumulator.cumulate(ctx.alloc(), cumulation,
                                                                data);
            }
            //从 cumulation 字节中解码出消息
            callDecode(ctx, cumulation, out);
        } catch (DecoderException e) {
            throw e;
        } catch (Exception e) {
            throw new DecoderException(e);
        } finally {
            /**
             * 当字节容器不为空且不可读时，需要释放
             * 并置空，直接回收，将下次解码认为是第一次
             */
            if (cumulation != null && !cumulation.isReadable()) {
                numReads = 0;
                cumulation.release();
```

```
                    cumulation = null;
                } else if (++ numReads >= discardAfterReads) {
                    //如果读取的字节数大于或等于discardAfterReads,
                    // 则设置读取字节数为0
                    // 并移除字节容器中一部分读取过的字节
                    numReads = 0;
                    discardSomeReadBytes();
                }
                int size = out.size();
                /**
                 * firedChannelRead 属性在 channelReadComplete()方法中被调用
                 */
                firedChannelRead |= out.insertSinceRecycled();
                //遍历解码消息集合,转发消息到下一个Handler处理器中
                fireChannelRead(ctx, out, size);
                //回收解码消息集合,以便下次循环利用
                out.recycle();
            }
        } else {
            //非ByteBuf消息,此解码器不进行解码
            ctx.fireChannelRead(msg);
        }
    }
}
protected void callDecode(ChannelHandlerContext ctx, ByteBuf in,
 List<Object> out) {
        try {
            //循环解码
            while (in.isReadable()) {
                int outSize = out.size();
                //判断是否已经有可用的消息
                if (outSize > 0) {
                    //触发下一个Handler去处理这些解码出来的消息
                    fireChannelRead(ctx, out, outSize);
                    out.clear();
                    //检测Handler是否被从通道处理器上下文移除了
                    //若被移除了,则不能继续操作
                    if (ctx.isRemoved()) {
```

```java
                break;
            }
            outSize = 0;
        }
        //获取字节容器的可读字节数
        int oldInputLength = in.readableBytes();
        //解码字节 buf 中的数据为消息对象,并将其放入 out 中
        // 如果解码器被从通道处理器上下文移除了,则处理移除事件
        decodeRemovalReentryProtection(ctx, in, out);
        if (ctx.isRemoved()) {
            break;
        }
        if (outSize == out.size()) {
            //如果可读字节数无变化,则说明解码失败,无须继续解码
            if (oldInputLength == in.readableBytes()) {
                break;
            } else {
                continue;
            }
        }
        //异常
        if (oldInputLength == in.readableBytes()) {
            throw new DecoderException(
                    StringUtil.simpleClassName(getClass()) +
                    ".decode() did not read anything … message.");
        }
        //是否只能解码一次
        if (isSingleDecode()) {
            break;
        }
    }
} catch (DecoderException e) {
    throw e;
} catch (Exception cause) {
    throw new DecoderException(cause);
}
}
```

```java
final void decodeRemovalReentryProtection(ChannelHandlerContext ctx,
                    ByteBuf in, List<Object> out)
        throws Exception {
    decodeState = STATE_CALLING_CHILD_DECODE;
    try {
        //由子类完成
        decode(ctx, in, out);
    } finally {
        //Channel 的处理器是否正在移除
        boolean removePending = decodeState
                    == STATE_HANDLER_REMOVED_PENDING;
        decodeState = STATE_INIT;
        if (removePending) {
            //处理 Handler 从通道处理器移除事件
            handlerRemoved(ctx);
        }
    }
}

public static final ByteToMessageDecoder.Cumulator MERGE_CUMULATOR = new
ByteToMessageDecoder.Cumulator() {
    @Override
    public ByteBuf cumulate(ByteBufAllocator alloc, ByteBuf cumulation,
    ByteBuf in) {
        try {
            final ByteBuf buffer;
            //判断是否需要扩容，其逻辑与组合缓冲区类似
            if (cumulation.writerIndex() > cumulation.maxCapacity()
                    - in.readableBytes()
                    || cumulation.refCnt() > 1 || cumulation.isReadOnly()) {
                buffer = expandCumulation(alloc, cumulation,
                in.readableBytes());
            } else {
                buffer = cumulation;
            }
            //把需要解码的字节写入读半包字节容器中
            buffer.writeBytes(in);
```

```
            return buffer;
        } finally {
            //非组合缓冲区，需要释放 buf
            in.release();
        }
    }
}
```

以上代码的整体逻辑如下。

（1）channelRead()方法首先会判断 msg 是否为 ByteBuf 类型，只有在是的情况下才会进行解码。这也是为什么将 StringDecoder 等 MessageToMessageCodec 解码器放在 ByteToMessageDecoder 子类解码器后面的原因，这时的 msg 一般是堆外直接内存 DirectByteBuf，因为采用堆外直接内存在传输时可以少一次复制。然后判断是否为第一次解码，若是，则直接把 msg 赋值给 cumulation （cumulation 是读半包字节容器）；若不是，则需要把 msg 写入 cumulation 中，写入之前要判断是否需要扩容。

（2）把新读取到的数据写入 cumulation 后，调用 callDecode()方法。在 callDecode()方法中会不断地调用子类的 decode()方法，直到当前 cumulation 无法继续解码。无法继续解码分两种情况：第一种情况是无可读字节；第二种情况是经历过 decode()方法后，可读字节数没有任何变化。

（3）执行完 callDecode()方法后，进入 finally 代码块进行收尾工作。若 cumulation 不为空，且不可读时，需要把 cumulation 释放掉并赋空值，若连续 16 次（discardAfterReads 的默认值）字节容器 cumulation 中仍然有未被业务拆包器读取的数据，则需要进行一次压缩：将有效数据段整体移到容器首部，同时用一个成员变量 firedChannelRead 来标识本次读取数据是否拆到了一个业务数据包，并触发 fireChannelRead 事件，将拆到的业务数据包传递给后续的 Handler，最后把 out 放回对象池中。

具体的解码器一般分两种：一种是根据特殊字符解码（如 DelimiterBasedFrameDecoder），

找到 ByteBuf 中是否有对应的特殊字符，若有，则截断读取对应的消息；另一种是根据写入消息体长度值解码（如 LengthFieldBasedFrameDecoder），这种解码器的一般用法是先读取前面 4 个字节的 int 值，再根据这个值去读取可用的数据包。

5.2.3　NioEventLoop 就绪处理之 OP_READ（二）

下面主要通过业务 Handler 返回结果的过程来详细讲解 Netty 的写处理逻辑。在具体分析之前，先思考以下几个问题。

第一个问题：Netty 在写操作时会依次调用哪些 Handler？

第二个问题：在业务 Handler 中，若开启了额外业务线程，那么在 Netty 内部是如何把业务线程的结果数据经过 I/O 线程发送出去的呢？

第三个问题：为了提高网络的吞吐量，在调用 write 时，数据并没有直接被写到 Socket 中，而是被写到了 Netty 的缓冲区（ChannelOutboundBuffer）中，在并发很大的情况下，当对方接收数据较慢时，Netty 的写缓冲区如何防止内存溢出，防止出现大量内存无法释放的情况。

第一个问题比较好解决，只需在 ServerHandler 的 channelRead()方法中调试，一步步跟进去即可找到。可以把这些处理类通过先后顺序串联起来，如图 5-6 所示。

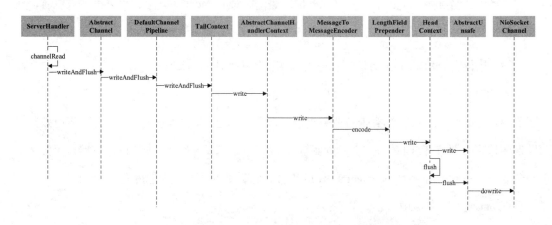

图 5-6　NioEventLoop 处理就绪 OP_READ 事件的时序图 2

可以将图 5-6 和图 5-5 进行简单的对比，在读数据时，Handler 从 HeadContext 开始到解码器父类 ByteToMessageDecoder，再到具体解码器，最后调用业务逻辑处理类 ServerHandler。在写数据时，Handler 从 TailContext 开始到编码器父类 MessageToMessageEncoder，再到具体编码器，最后调用 HeadContext。其中的 TailContext 只起了引用串联的作用，具体逻辑处理是由其父类实现的。而 HeadContext 的一个属性——unsafe，此属性处理了 Channel 的接入和数据的写入、关闭等。Handler 链的顺序如图 5-7 所示。

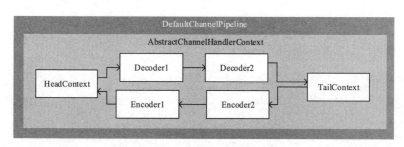

图 5-7 Handler 链的顺序

图 5-7 中的 HeadContext 和 TailContext 之间连接了各种编码器、解码器，形成了整个 Handler 链表，图中的箭头表示编码器和解码器的查找方向。Handler 链表的头部和尾部都是在 DefaultChannelPipeline 的构造方法中定义好的，具体代码如下：

```
protected DefaultChannelPipeline(Channel channel) {
    this.channel = ObjectUtil.checkNotNull(channel, "channel");
    succeededFuture = new SucceededChannelFuture(channel, null);
    voidPromise = new VoidChannelPromise(channel, true);
    tail = new TailContext(this);
    head = new HeadContext(this);
    head.next = tail;
    tail.prev = head;
}
```

通过 addLast() 方法追加进去的编码器和解码器都位于 TailContext 的前面。

```
private void addLast0(AbstractChannelHandlerContext newCtx) {
    AbstractChannelHandlerContext prev = tail.prev;
    newCtx.prev = prev;
```

```
    newCtx.next = tail;
    prev.next = newCtx;
    tail.prev = newCtx;
}
```

当发生读事件（在 Netty 里也叫输入 inbound 事件）时，I/O EventLoop 线程先从 HeadContext 中依次向后查找 ChannelInboundHandler 类型的 Handler，并调用其 channelRead() 方法。当发生写操作 outbound 事件时，从 TailContext 中依次向前查找 ChannelOutboundHandler 类型的 Handler，并调用其 write() 方法。这也是为何解码器先追加的被先调用，而编码器正好相反的缘故。下面是对 Netty 分别查找 inbound() 和 outbound() 方法的解读：

```
private AbstractChannelHandlerContext findContextInbound(int mask) {
    AbstractChannelHandlerContext ctx = this;
    do {
        ctx = ctx.next;
    } while ((ctx.executionMask & mask) == 0);
    return ctx;
}
private AbstractChannelHandlerContext findContextOutbound(int mask) {
    AbstractChannelHandlerContext ctx = this;
    do {
        ctx = ctx.prev;
    } while ((ctx.executionMask & mask) == 0);
    return ctx;
}
```

至此，第一个问题告一段落，由于编码的细节比较简单，此处不再细说。再来看第二个问题，在 ServerHandler 里开启额外线程去执行 ctx.channel().writeAndFlush(JSONObject.toJSONString(response)) 时，NioEventLoop 线程如何获取 response 内容并写回给 Channel 呢？

在写的过程中有两种 task，分别是 WriteTask 和 WriteAndFlushTask，主要根据是否刷新来决定使用哪种 task。在 NioSocketChannel 中，每个 Channel 都有一条 NioEventLoop 线程与之对应，在 NioEventLoop 的父类 SingleThreadEventExecutor 中有个队列属性，叫 taskQueue，它主要通过 SingleThreadEventExecutor 的 execute() 方法存放非 EventLoop 线程的任务，包括

WriteTask 和 WriteAddFlushTask 这两种 WriteTask。当调用添加任务时，会唤醒 EventLoop 线程，从而 I/O 线程会去调用这些任务的 run()方法，并把结果写回 Socket 通道。具体核心代码如下：

```
    EventExecutor executor = next.executor();
    if (executor.inEventLoop()) {//判断是否为 EventLoop 线程
        if (flush) {
            next.invokeWriteAndFlush(m, promise);
        } else {
            next.invokeWrite(m, promise);
        }
} else {//当为非 EventLoop 线程时需要构建 task
    final AbstractWriteTask task;
    if (flush) {
        task = WriteAndFlushTask.newInstance(next, m, promise);
    } else {
        task = WriteTask.newInstance(next, m, promise);
    }
    //把 task 加入 executor 中，这个 executor 就是 NioEventLoop
    //若加入失败，则取消 task 的执行
    if (!safeExecute(executor, task, promise, m)) {
            task.cancel();
    }
```

在解决第三个问题前，先来了解一下 Netty 的缓冲区 ChannelOutboundBuffer。ChannelOutboundBuffer 由一个链表构成，链表的 Entry 中有消息内容、next 指针等，其中有 5 个非常重要的属性：

```
//链表中被刷新的第一个元素，此元素准备第一个写入 Socket
private Entry flushedEntry;
//链表中第一个未刷新的元素
//当调用 addMessage()方法后，从原链表 tailEntry 到 Entry（现链表的 tailEntry）节点
//都是未被刷新的数据
private Entry unflushedEntry;
//链表末尾节点
private Entry tailEntry;
//表示已经刷新但还没有写到 Socket 中的 Entry 的数量
```

```
private int flushed;
// 通道缓存总数据，用来控制背压
//每新增一个 Entry，其大小要加上 Entry 实例的大小（96B）和真实数据的大小
private volatile long totalPendingSize;
```

前面 3 个属性都只是指针，构建了刷新和未刷新的数据链表。ChannelOutboundBuffer 缓冲区处理过程如图 5-8 所示。

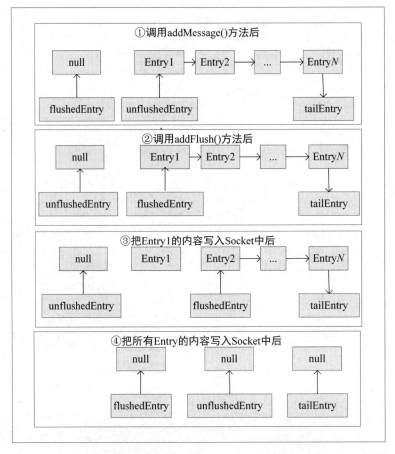

图 5-8　ChannelOutboundBuffer 缓冲区处理过程

在调用 addMessage()方法后，采用 CAS 方式增加待发送节点的字节数，此时如果待发送的字节数大于通道写 buf 的最高阈值 writeBufferHighWaterMark，则更新通道状态为不可写，同时会触发 channelWritabilityChanged 事件，防止内存溢出。至于在 ServerHandler 的 channelWritabilityChanged()

方法中进行怎样的处理，读者可回到 2.6 节仔细看看 Flink 的背压。

ChannelOutboundBuffer 的代码解读顺序与图 5-8 的处理过程一致，从 addMessage()方法添加消息到缓存区开始，调用 addFlush()方法标识待发送的 Entry，并调用 AbstractUnsafe 的 flush0，最终到 NioSocketChannel 的 doWrite()方法，才真正地把所有 Entry 的内容写入了 Socket 中。其中，flush0()方法与 doWrite()方法在第 4 章已经进行了详细的剖析，此处不再描述。下面只剖析与 ChannelOutboundBuffer 相关的部分，包括 ChannelOutboundBuffer 的 nioBuffers()、removeBytes()与 remove()方法。

addMessage()方法的代码解读如下：

```
/**
 * 每调用一次 ctx.channel().write
 * 最终都会触发 addMessage()方法，并把数据加在 tailEntry 后面
 * @param msg
 * @param size
 * @param promise
 */
public void addMessage(Object msg, int size, ChannelPromise promise) {
    //把 msg 消息数据包装成 Entry 对象
    ChannelOutboundBuffer.Entry entry = ChannelOutboundBuffer.Entry.
                                    newInstance(msg, size, total(msg), promise);
    //若链表为空，则尾节点为当前节点
    if (tailEntry == null) {
        flushedEntry = null;
    } else {
        //当链表不为空时，把新的 Entry 对象添加到链表尾部
        ChannelOutboundBuffer.Entry tail = tailEntry;
        tail.next = entry;
    }
    tailEntry = entry;
    /**
     * unflushedEntry 为空，
     * 表示调用 addFlush()方法将链表中之前的元素都已经全部加入了需要发送的节点
     * 否则链表为空
```

```java
     */
    if (unflushedEntry == null) {
        unflushedEntry = entry;
    }
    /**
     * 修改通道缓存总数据的大小，若缓存总数据大小超过了高水位
     * 则会触发 fireChannelWritabilityChanged 事件，进入背压
     */
    incrementPendingOutboundBytes(entry.pendingSize, false);
}
```

addFlush()方法用来修改缓冲区中的数据状态，它会提前把消息标识为已发送，在标识时，需要注意写水位和重要属性的变化。具体代码解读如下：

```java
/**
 * 移动 unflushedEntry 的位置，修改待发送 Entry 的个数
 * 设置每个 entry 的状态为非取消状态
 */
public void addFlush() {
    ChannelOutboundBuffer.Entry entry = unflushedEntry;
    if (entry != null) {
        if (flushedEntry == null) {
            // there is no flushedEntry yet, so start with the entry
            flushedEntry = entry;
        }
        //从 unflushedEntry 开始循环设置，将这些 entry 状态设置为非取消状态
        do {
            flushed ++;
            if (!entry.promise.setUncancellable()) {
                int pending = entry.cancel();
                /**
                 * entry 如果已经取消
                 * 则释放 entry 对应的内存
                 * 减小 ChannelOutboundBuffer 的大小
                 * 如果缓存总数据的大小低于低水位
                 * 则触发 fireChannelWritabilityChanged 事件
                 * 调用业务 Handler 的 channelWritabilityChanged() 方法
                 */
```

```
                decrementPendingOutboundBytes(pending, false, true);
            }
            entry = entry.next;
    } while (entry != null);
    /**
     * 每次设置完后都需要把 unflushedEntry 置空
     * 在下次添加数据时,unflushedEntry 为最先添加的 entry
     */
    unflushedEntry = null;
    }
}
```

nioBuffers()方法把缓冲区中需要发送的数据转换成了 ByteBuffer,因为 NIO 的 SocketChannel 只能写 ByteBuffer 类型的数据。具体代码解读如下:

```
/**
 * 在发送数据时需要把 ChannelOutboundBuffer 中的 msg 转换成 ByteBuffer
 * @param maxCount 本次最多获取 buf 的个数为 1024
 * @param maxBytes 本次获取最大字节数
 * @return
 */
public ByteBuffer[] nioBuffers(int maxCount, long maxBytes) {
    assert maxCount > 0;
    assert maxBytes > 0;
    long nioBufferSize = 0;
    int nioBufferCount = 0;
    final InternalThreadLocalMap threadLocalMap =
                                InternalThreadLocalMap.get();
    //从线程本地缓存中获取 ByteBuffer 数组
    ByteBuffer[] nioBuffers = NIO_BUFFERS.get(threadLocalMap);
    //从准备第一个写入 Socket 的元素开始
    ChannelOutboundBuffer.Entry entry = flushedEntry;
    /**
     * 循环遍历 entry
     * entry 必须为准备写入 Socket 的元素且为非取消状态
     */
    while (isFlushedEntry(entry) && entry.msg instanceof ByteBuf) {
        if (!entry.cancelled) {
```

```java
//获取entry节点中实际发送的数据
ByteBuf buf = (ByteBuf) entry.msg;
final int readerIndex = buf.readerIndex();
//获取可发送字节数
final int readableBytes = buf.writerIndex() - readerIndex;
//若可发送字节数大于0则继续；否则跳过
if (readableBytes > 0) {
    /**
     * 累计发送字节数不能大于maxBytes
     */
    if (maxBytes - readableBytes < nioBufferSize &&
            nioBufferCount != 0) {
        break;
    }
    //累计发送字节数
    nioBufferSize += readableBytes;
    //获取节点中ByteBuffer的个数
    int count = entry.count;
    if (count == -1) {
        //noinspection ConstantValueVariableUse
        entry.count = count = buf.nioBufferCount();
    }
    //需要存放多少个ByteBuffer
    int neededSpace = min(maxCount, nioBufferCount + count);
    //nioBuffers长度不够，需要扩容
    if (neededSpace > nioBuffers.length) {
        nioBuffers = expandNioBufferArray(nioBuffers,
                        neededSpace, nioBufferCount);
        NIO_BUFFERS.set(threadLocalMap, nioBuffers);
    }
    //如果ByteBuffer的个数为1，则直接获取ByteBuffer并放入nioBuffers数组中
    if (count == 1) {
        ByteBuffer nioBuf = entry.buf;
        if (nioBuf == null) {
            entry.buf = nioBuf = buf.internalNioBuffer(readerIndex,
                            readableBytes);
        }
```

```
                    nioBuffers[nioBufferCount++] = nioBuf;
                } else {
                    //如果有多个循环获取ByteBuffer放入nioBuffers数组中
                    nioBufferCount = nioBuffers(entry, buf, nioBuffers,
                                    nioBufferCount, maxCount);
                }
                //不能超过最大个数限制
                if (nioBufferCount == maxCount) {
                    break;
                }
            }
        }
        //获取下一个节点
        entry = entry.next;
    }
    this.nioBufferCount = nioBufferCount;
    this.nioBufferSize = nioBufferSize;
    return nioBuffers;
}
```

通过 nioBuffers() 方法获取到需要发送的 ByteBuffer 数组，然后通过 SocketChannel 写到网络中，并返回写成功了多少个字节，此时 ChannelOutboundBuffer 需要把这些字节从链表中移除。同时需要把刚刚生成的 ByteBuffer 数组也一起移除，下面继续看 removeBytes() 与 remove() 方法的解读：

```
/**
 * 移除写成功的字节
 * @param writtenBytes 字节数
 */
public void removeBytes(long writtenBytes) {
    for (;;) {
        /**
         * 与 nioBuffers() 方法一样，从准备写入 Socket 的节点开始
         * 获取此节点的 buf 数据
         */
        Object msg = current();
        if (!(msg instanceof ByteBuf)) {
```

```
            assert writtenBytes == 0;
            break;
        }
        final ByteBuf buf = (ByteBuf) msg;
        final int readerIndex = buf.readerIndex();
        //获取 buf 可发送字节数
        final int readableBytes = buf.writerIndex() - readerIndex;
        /**
         * 如果当前节点的字节数小于或等于已发送的字节数
         * 则直接删除整个节点,并更新进度
         */
        if (readableBytes <= writtenBytes) {
            if (writtenBytes != 0) {
                progress(readableBytes);
                writtenBytes -= readableBytes;
            }
            remove();
        } else { //若当前节点还有一部分未发送,则缩小当前节点的可发送字节长度
            if (writtenBytes != 0) {
                //修改其 readerIndex 并更新进度
                buf.readerIndex(readerIndex + (int) writtenBytes);
                progress(writtenBytes);
            }
            break;
        }
    }
    /**
     * 由于每次在发送时
     * 都需要从线程本地缓存中获取 ByteBuffer 数组
     * 且每次获取的数组应无任何数据
     * 因此此处需要清空它
     */
    clearNioBuffers();
}
private void clearNioBuffers() {
    int count = nioBufferCount;
    if (count > 0) {
```

```java
        nioBufferCount = 0;
        //填null对象
        Arrays.fill(NIO_BUFFERS.get(), 0, count, null);
    }
}
/**
 * 节点数据都发送完后
 * 需要把此节点从链表中移除
 * @return
 */
public boolean remove() {
    //与nioBuffers()、removeBytes()方法一样
    ChannelOutboundBuffer.Entry e = flushedEntry;
    if (e == null) {
        //如果获取不到链头节点,则清空ByteBuf缓存
        clearNioBuffers();
        return false;
    }
    Object msg = e.msg;
    ChannelPromise promise = e.promise;
    int size = e.pendingSize;
    //从链表中移除此节点,同时将flushedEntry指针指向下一节点
    removeEntry(e);
    if (!e.cancelled) {
        /**
         * 节点在非取消状态下
         * 由于没有地方用得上节点数据,因此需要释放其内存空间
         * 并通知处理成功
         * 同时缓存总数据大小相应的减小
         */
        ReferenceCountUtil.safeRelease(msg);
        safeSuccess(promise);
        decrementPendingOutboundBytes(size, false, true);
    }
    //回收Entry对象并放回对象池
    e.recycle();
    return true;
```

```
}
/**
 * 移除节点
 * 同时修改 flushedEntry 指针
 * @param e
 */
private void removeEntry(ChannelOutboundBuffer.Entry e) {
    if (-- flushed == 0) {
        //若最后的节点也被移除了,则所有指针为 null
        flushedEntry = null;
        if (e == tailEntry) {
            tailEntry = null;
            unflushedEntry = null;
        }
    } else {
        //否则预写入指针会不断向前移动
        flushedEntry = e.next;
    }
}
```

通过本章的学习,知道了 Netty 内部是如何接收请求获取请求数据,并把结果写回给请求方的。希望读者反复思考,根据分布式 RPC 实例进行反复调试。

第 6 章

Netty 内存管理

6.1 Netty 内存管理策略介绍

为了提高内存的使用效率，Netty 引入了 jemalloc 内存分配算法。Netty 内存管理层级结构如图 6-1 所示，其中右边是内存管理的 3 个层级，分别是本地线程缓存、分配区 arena、系统内存；左边是内存块区域，不同大小的内存块对应不同的分配区，总体的分配策略如下。

- 为了避免线程间锁的竞争和同步，每个 I/O 线程都对应一个 PoolThreadCache，负责当前线程使用非大内存的快速申请和释放。
- 当从 PoolThreadCache 中获取不到内存时，就从 PoolArena 的内存池中分配。当内存使用完并释放时，会将其放到 PoolThreadCache 中，方便下次使用。若从 PoolArena 的内存池中分配不到内存，则从堆内外内存中申请，申请到的内存叫 PoolChunk。当内存块的大小默认为 12MB 时，会被放入 PoolArea 的内存池中，以便重复利用。当申请大内存时（超过了 PoolChunk 的默认内存大小 12MB），直接在堆外或堆内内存中创建（不归 PoolArea 管理），用完后直接回收。本书只介绍 Netty 的内存池可重复利用的内存。

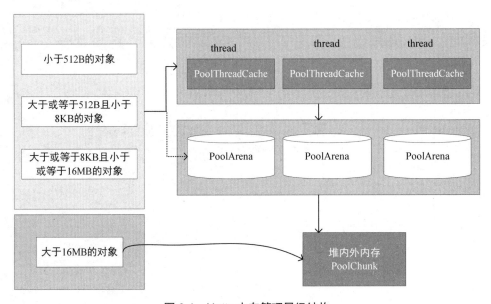

图 6-1 Netty 内存管理层级结构

Netty 内存分配思维导图如图 6-2 所示。

（1）Netty 在具体分配内存之前，会先获取本次内存分配的大小。具体的内存分配由 PoolArena 统一管理，先从线程本地缓存 PoolThreadCache 中获取，线程本地缓存采用固定长度队列缓存此线程之前用过的内存。

（2）若本地线程无缓存,则判断本次需要分配的内存大小,若小于 512B,则先从 PoolArena 的 tinySubpagePools 缓存中获取；若大于或等于 512B 且小于 8KB，则先从 smallSubpagePools 缓存中获取,上述两种情况缓存的对象都是 PoolChunk 分配的 PoolSubpage；若大于或等于 8KB 或在 SubpagePools 缓存中分配失败，则从 PoolChunkList 中查找可分配的 PoolChunk。

（3）若 PoolChunkList 分配失败，则创建新的 PoolChunk，由 PoolChunk 完成具体的分配工作，最终分配成功后，加入对应的 PoolChunkList 中。若分配的是小于 8KB 的内存，则需要把从 PoolChunk 中分配的 PoolSubpage 加入 PoolArena 的 SubpagePools 中。

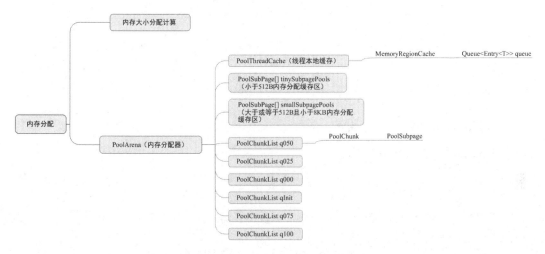

图 6-2　Netty 内存分配思维导图

6.2　PoolChunk 内存分配

6.2.1　PoolChunk 分配大于或等于 8KB 的内存

Netty 底层的内存分配和管理主要由 PoolChunk 实现，大于 16MB 的 PoolChunk 由于不放入内存池管理，比较简单，这里不进行过多的讲解。下面主要讲解 PoolChunk 为何能管理内存，以及它具有的重要属性。

可以把 Chunk 看作一块大的内存，这块内存被分成了很多小块的内存，Netty 在使用内存时，会用到其中一块或多块小内存。如图 6-3 所示，通过 0 和 1 来标识每位的占用情况，通过内存的偏移量和请求内存空间大小 reqCapacity 来决定读/写指针。在图 6-3 中，浅灰色部分表示内存实际用到的部分，深灰色部分表示浪费的内存。内存池在分配内存时，只会预先准备固定大小和数量的内存块，不会请求多少内存就恰好给多少内存，因此会有一定的内存被浪费。

使用完后交还给 PoolChunk 并还原，以便重复使用。

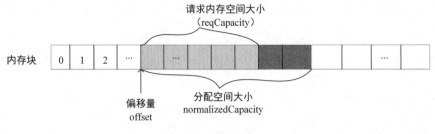

图 6-3　内存块运用

PoolChunk 内部维护了一棵平衡二叉树，默认由 2048 个 page 组成，一个 page 默认为 8KB，整个 Chunk 默认为 16MB，其二叉树结构如图 6-4 所示。

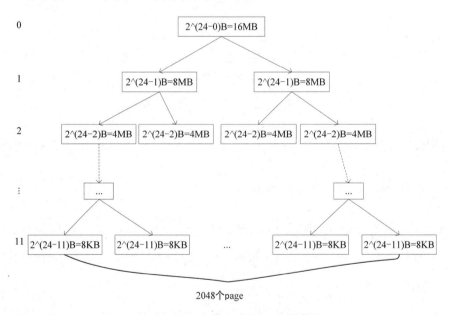

图 6-4　PoolChunck 的二叉树结构

当分配的内存大于 $2^{13}B$（8196B）时，可以通过内存值计算对应的层级：int d = 11-(log2(normCapacity)-13)。其中，normCapacity 为分配的内存大小，它大于或等于 8KB 且为 8KB 的整数倍。例如，申请大小为 16KB 的内存，d=11-(14-13)=10，表示只能在小于或等于 10 层上寻找还未被分配的节点。

在 PoolChunk 中，用一个数组 memoryMap 维护了所有节点（节点数为 1～2048×2-1）及其对应的高度值。memoryMap 是在 PoolChunk 初始化时构建的，其下标为图 6-4 中节点的位置，其值为节点的高度，如 memoryMap[1]=0、memoryMap[2]=1、memoryMap[2048]=11、memoryMap[4091]=11。另外，还有一个同样的数组——depthMap。两者的区别是：depthMap 一直不会改变，通过 depthMap 可以获取节点的内存大小，还可以获取节点的初始高度值；而 memoryMap 的节点和父节点对应的高度值会随着节点内存的分配发生变化。当节点被全部分配完时，它的高度值会变成 12，表示目前已被占用，不可再被分配，并且会循环递归地更新其上所有父节点的高度值，高度值都会加 1，如图 6-5 所示。

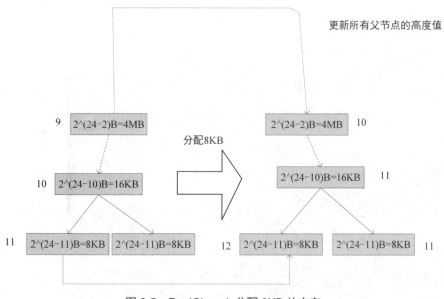

图 6-5　PoolChunck 分配 8KB 的内存

Netty 源码是如何查找对应的可用节点并更新其父节点的高度值的呢？

Netty 采用了前序遍历算法，从根节点开始，第二层为左右节点，先看左边节点内存是否够分配，若不够，则选择其兄弟节点（右节点）；若当前左节点够分配，则需要继续向下一层层地查找，直到找到层级最接近 d（分配的内存在二叉树中对应的层级）的节点。具体查找算法如下：

```
//d 是申请的内存在 PoolChunck 二叉树中的高度值,若内存为 8KB,则 d 为 11
private int allocateNode(int d) {
int id = 1;
int initial = - (1 << d);//掩码,与id进行与操作后,若>0,则说明id对应的高度大于或等于d
byte val = value(id);// 为 memoryMap[id]
if (val > d) { // 若当前分配的空间无法满足要求,则直接返回-1,分配失败
    return -1;
}
/**
有空间可以分配了,就需要一步步找到更接近高度值为d的节点 ,若找到的高度值等于d
但此时其下标与 initial 进行与操作后为 0,则说明其子节点有一个未被分配,且其初始化层级<d
只是由于其有一个节点被分配了,所以层级 val 与 d 相等
*/
while (val < d || (id & initial) == 0) {
    //每次都需要把id向下移动一层,即左移一位
    id <<= 1;
    //获取id对应的层级高度值 memoryMap[id]
    val = value(id);
    //若id对应层级的高度值大于d,则此时去其兄弟节点找,肯定能找到
    if (val > d) {
        //获取其兄弟节点,兄弟节点的位置通过id值异或1得到
        id ^= 1;
        //获取其兄弟节点的高度值
        val = value(id);
    }
}
byte value = value(id);//获取找到的节点的高度值
setValue(id, unusable);  //标识为不可用
updateParentsAlloc(id);//更新其所有父节点的高度值
return id;//返回 id (1~2048*2-1),通过 id 可以获取其层级高度值
        //也可以算出其占用的内存空间的大小
}
```

通过上述图、文字及代码,描述了 Netty 在分配[8KB,16MB]内存时是采用怎样的方式快速找到对应可用的内存块的 id 的,此 id 为 PoolChunk 的内存映射属性 memoryMap 的下标,通过此下标可以找到其偏移量。获取偏移量的算法为:int shift = id ^ 1 << depth(id);int offset =

shift*runLength(id);。主要通过 id 异或与 id 同层级最左边的元素的下标值得到偏移量，再用偏移量乘以当前层级节点的内存大小，进而获取在 PoolChunk 整个内存中的偏移量。有了偏移量和需要分配的内存大小 Length，以及最大可分配内存的大小（可根据 runLength(id) 计算得出），即可初始化 PooledByteBuf，完成内存分配。runLength() 方法为 return 1 << 24 - depth(id)，其中 id 的层级越大，返回的值越小，当 id 为 0 时，返回 16MB，为整个 PoolChunk 内存的大小；当 id 为 11 时，返回 8KB，为 page 默认的大小。

肯定有读者会感到疑惑：若需要分配的内存不在 memoryMap 数组里该怎么办呢？此时无法通过 int d = 11-(log2(normCapacity)-13) 来计算。然而在内存具体分配之前，Netty 对需要的内存量进行了加工处理（会根据其大小进行处理），若需要的内存大于或等于 512B，则此时计算一个大于且距它最近的 2 的指数次幂的值，这个值就是 PooledByteBuf 的最大可分配内存。若小于 512B，则找到距 reqCapacity 最近的下一个 16 的倍数的值。

大于或等于 page 的内存分配基本上讲完了，已详细了解了其内存的具体分配算法。但在实际应用中，一般接收的请求信息都不大，很多时候请求数据不会超过 8KB。在小于 8KB 的情况下，Netty 的内存分配又是怎样的呢？下面通过处理 PoolSubpage 的逻辑来解决这个问题。

6.2.2　PoolChunk 分配小于 8KB 的内存

6.2.1 小节详细了解了 PoolChunk 分配大于或等于 8KB 的内存，那么对于内存小于 8KB 的情况，Netty 又是如何设计的呢？PoolChunk 有一个 PoolSubpage[] 组数，叫 subpages，它的数组长度与 PoolChunk 的 page 节点数一致，都是 2048，由于它只分配小于 8KB 的数据，且 PoolChunk 的每个 page 节点只能分配一种 PoolSubpage，因此 subpages 的下标与 PoolChunk 的 page 节点的偏移位置是一一对应的。PoolChunk 上的每个 page 均可分配一个 PoolSubpage 链表，由于链表中元素的大小都相同，因此这个链表最大的元素个数也已经确定了，链表的头节点为 PoolSubpage[] 组数中的元素。PoolChunk 分配 PoolSubpage 的步骤如下。

步骤 1：在 PoolChunk 的二叉树上找到匹配的节点，由于小于 8KB，因此只匹配 page 节点。

因为 page 节点从 2048 开始，所以 page 将节点与 2048 进行异或操作就可以得到 PoolSubpage[] 数组 subpages 的下标 subpageIdx。

步骤 2：取出 subpages[subpageIdx]的值。若值为空，则新建一个 PoolSubpage 并初始化。

步骤 3：PoolSubpage 根据内存大小分两种，即小于 512B 为 tiny，大于或等于 512B 且小于 8KB 为 small，此时 PoolSubpage 还未区分是 small 还是 tiny。然后把 PoolSubpage 加入 PoolArena 的 PoolSubpage 缓存池中，以便后续直接从缓存池中获取。

步骤 4：PoolArena 的 PoolSubpage[]数组缓存池有两种，分别是存储(0,512)个字节的 tinySubpagePools 和存储[512,8192)个字节的 smallSubpagePools。由于所有内存分配大小 elemSize 都会经过 normalizeCapacity 的处理，因此当 elemSize>=512 时，elemSize 会成倍增长，即 512→1024→2048→4096→8192；当 elemSize<512 时，elemSize 从 16 开始，每次加 16B，elemSize 的变化规律与 tinySubpagePools 和 smallSubpagePools 两个数组的存储原理是一致的。

tinySubpagePools 从 16 开始存储，根据下标值加 16B，因此其数组长度为 512/16B=32B。smallSubpagePools 从 512 开始，根据下标值成倍地增长。数组总共有 4 个元素。通过 elemSize 可以快速找到合适的 poolSubpage。

tinysubpage 通过 elemSize>>>4 即可获取 tinysubpage[]数组的下标 tableIdx，而 smallSubpage 则需要除 1024 看是否为 0，若不为 0，则通过循环除以 2 再看是否为 0，最终找到 tableIdx。有了 tableIdx 就可以获取对应的 PoolSubpage 链表，若 PoolSubpage 链表获取失败，或链表中不可再分配元素，则回到步骤 1，从 PoolChunk 中重新分配；否则直接获取链表中可分配的元素。

根据以上逻辑分析，可以用图形方式来描述 PoolChunk 中 page 与 PoolSubpage 和 Arena 中两个 PoolSubpages 的对应关系，如图 6-6 所示。

第 6 章 Netty 内存管理

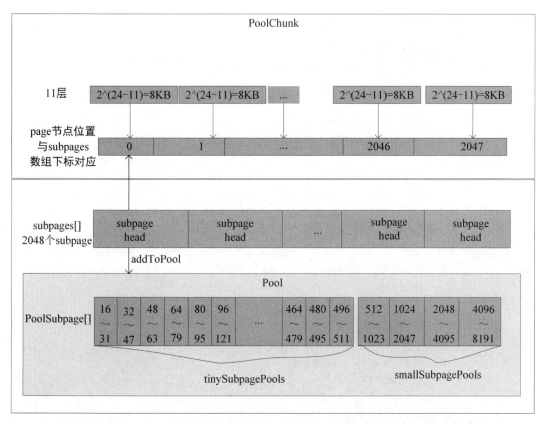

图 6-6 PoolChunck 分配 subpage 内存

PoolChunk 分配小于 8KB 内存的具体代码解读如下：

```
//当申请内存小于 8KB 时，此方法被调用
private long allocateSubpage(int normCapacity) {
//通过优化后的内存容量找到 Arena 的两个 subpages 缓存池其中一个的对应空间 head 指针
PoolSubpage<T> head = arena.findSubpagePoolHead(normCapacity);
int d = maxOrder;//小于 8KB 内存只在 11 层分配
//由于分配前需要把 PoolSubpage 加入缓存池中，以便下回直接从 Arena 的缓存池中获取
//因此选择加锁 head 指针
synchronized (head) {
    //获取一个可用的节点
    int id = allocateNode(d);
    if (id < 0) {
        return id;
```

```java
        }
        final PoolSubpage<T>[] subpages = this.subpages;
        final int pageSize = this.pageSize;
        //可用空间减去一个 page，表示此 page 被占用
        freeBytes -= pageSize;
        //根据 page 的偏移值减 2048 获取 PoolSubpage 的索引
        int subpageIdx = subpageIdx(id);
        //获取 page 对应的 PoolSubpage
        PoolSubpage<T> subpage = subpages[subpageIdx];
        if (subpage == null) {
            //若为空，则初始化一个，初始化会运行 PoolSubpage 的 addToPool()方法
            //把 subpage 追加到 head 的后面
            subpage = new PoolSubpage<T>(head, this, id, runOffset(id), pageSize,
                                        normCapacity);
            subpages[subpageIdx] = subpage;
        } else {
            //初始化同样会调用 addToPool()方法
            //此处思考，什么情况下才会发生这类情况
            subpage.init(head, normCapacity);
        }
    //PoolSubpage 的内存分配
    return subpage.allocate();
    }
}
PoolSubpage<T> findSubpagePoolHead(int elemSize) {
    int tableIdx;
    PoolSubpage<T>[] table;
    if (isTiny(elemSize)) { //判断是否 < 512
        tableIdx = elemSize >>> 4;//除以 16 即可
        table = tinySubpagePools;
    } else {
        tableIdx = 0;
        elemSize >>>= 10;//除以 1024
        while (elemSize != 0) { //elemSize 大于或等于 1024
            elemSize >>>= 1;//除以 2，因为后续字节都是以 2 为倍数来增长的
            tableIdx ++;
        }
```

```
        table = smallSubpagePools;
    }
    return table[tableIdx];
}
```

6.3 PoolSubpage 内存分配与释放

PoolSubpage 是由 PoolChunk 的 page 生成的，page 可以生成多种 PoolSubpage，但一个 page 只能生成其中一种 PoolSubpage。PoolSubpage 可以分为很多段，每段的大小相同，且由申请的内存大小决定。在讲解 PoolSubpage 具体分配内存之前，先看看它的重要属性：

```
final class PoolSubpage<T> implements PoolSubpageMetric {
final PoolChunk<T> chunk;//当前分配内存的 chunk
private final int memoryMapIdx;//当前 page 在 chunk 的 memoryMap 中的下标 id
private final int runOffset;//当前 page 在 chunk 的 memory 上的偏移量
private final int pageSize;//page 的大小，默认是 8192
private final long[] bitmap;//poolSubpage 每段内存的占用状态，采用二进制位来标识
PoolSubpage<T> prev;//指向前一个 PoolSubpage
PoolSubpage<T> next;//指向后一个 PoolSubpage
int elemSize;//切分后每段的大小
private int maxNumElems;//段的总数量
//实际采用二进制位标识的 long 数组的长度值，根据每段大小 elemSize 和 pageSize 计算得来
private int bitmapLength;
private int nextAvail;//下一个可用的位置
private int numAvail;//可用的段的数量
```

由于 PoolSubpage 每段的最小值为 16B，因此它的段的总数量最多为 pageSize/16。把 PoolSubpage 中每段的内存使用情况用一个 long[]数组来标识，long 类型的存储位数最大为 64B，每一位用 0 表示为空闲状态，用 1 表示被占用，这个数组的长度为 pageSize/16/64B，默认情况下，long 数组的长度最大为 8192/16/64B=8B。每次在分配内存时，只需查找可分配二进制的位置，即可找到内存在 page 中的相对偏移量，图 6-7 为 PoolSubpag 的内存段分配实例。

PoolSubpage 最多有8196/16=512段，需要用8个long类型的数字来描述这64个
内存段的状态，1表示被占用，0表示空闲

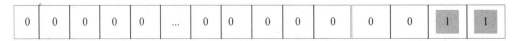

图 6-7　PoolSubpage 的内存段分配实例

PoolSubpage 的内存分配具体源码解读如下：

```
long allocate() {
if (elemSize == 0) {
    return toHandle(0);
}
if (numAvail == 0 || !doNotDestroy) {
    return -1;
}
//获取 PoolSubpage 下一个可用的位置
final int bitmapIdx = getNextAvail();
int q = bitmapIdx >>> 6;//获取该位置在 bitmap 数组对应的下标值
int r = bitmapIdx & 63;//获取 bitmap[q] 上实际可用的位
assert (bitmap[q] >>> r & 1) == 0;//确定该位没有被占用
bitmap[q] |= 1L << r;//将该位设为1，表示已被占用，此处 1L<<r 表示将 r 位设为 1
//若没有可用的段，则说明此 page/PoolSubpage 已经分配满了，没有必要继续放到 PoolArena 池中
//应从 pool 中移除
if (-- numAvail == 0) {
  removeFromPool();
}
//把当前 page 的索引和 PoolSubPage 的索引一起返回
//低32位表示 page 的 index, 高32位表示 PoolSubPage 的 index
return toHandle(bitmapIdx);
}
//查找下一个可用的位置
private int getNextAvail() {
int nextAvail = this.nextAvail;
//若下一个可用位置大于或等于0，则说明是第一次分配或正好已经有内存回收，可直接返回
if (nextAvail >= 0) {
    this.nextAvail = -1;//每次分配完内存后，都要将 nextAvail 设置为-1
    return nextAvail;
```

```java
    }
    //没有直接可用内存，需要继续查找
    return findNextAvail();
}
private int findNextAvail() {
    final long[] bitmap = this.bitmap;
    final int bitmapLength = this.bitmapLength;
    //遍历用来标识内存是否被占用的数组
    for (int i = 0; i < bitmapLength; i ++) {
        long bits = bitmap[i];
        //若当前 long 型的标识位不全为 1，则表示其中有未被使用的内存
        if (~bits != 0) {
            return findNextAvail0(i, bits);
        }
    }
    return -1;
}
private int findNextAvail0(int i, long bits) {
    final int maxNumElems = this.maxNumElems;
    //i 表示 bitmap 的位置，由于 bitmap 每个值有 64 位
    //因此用 i*64 来表示 bitmap[i]的第一位在 PoolSubpage 中的偏移量
    final int baseVal = i << 6;
    for (int j = 0; j < 64; j ++) {
        //判断第一位是否为 0，为 0 表示该位空闲
        if ((bits & 1) == 0) {
            //i*64|未被占用位，得到空位在 PoolSubpage 上的位置
            int val = baseVal | j;
            if (val < maxNumElems) {
                return val;
            } else {
                break;
            }
        }
        //若 bits 的第一位不为 0，则继续右移一位，判断第二位
        bits >>>= 1;
    }
    //若没有找到，则返回-1
```

```
    return -1;
}
//把当前 page 的索引和 PoolSubpage 的索引一起返回
//低 32 位表示 page 的 index, 高 32 位表示 PoolSubpage 的 index
private long toHandle(int bitmapIdx) {
    return 0x4000000000000000L | (long) bitmapIdx << 32 | memoryMapIdx;
}
```

通过上述代码解读可知，当内存空间在 PoolSubpage 中分配成功后，可以得到一个指针 handle。而 Netty 上层只会运用 ByteBuf，那么 ByteBuf 是如何跟 handle 关联的呢？

通过 handler 可以计算 page 的偏移量，也可以计算 subpage 的段在 page 中的相对偏移量，两者加起来就是该段分配的内存在 chunk 中的相对位置偏移量。当 PoolByteBuf 从 Channel 中读取数据时，需要用这个相对位置偏移量来获取 ByteBuffer。PoolByteBuf 从原 Buffer 池中通过 duplicate() 方法在不干扰原来 Buffer 索引的情况下，与其共享缓冲区数据，复制一份新的 ByteBuffer，并初始化新 ByteBuffer 的元数据，通过 offset 指定读/写位置，用 limit 来限制读/写范围。

PooledByteBuf 关键代码解读如下：

```
public final int setBytes(int index, ScatteringByteChannel in, int length){
    try {
        //从 NioSocketChannel 中读取请求数据，并将其写入 PoolByteBuf 中
        return in.read(internalNioBuffer(index, length));
    } catch (ClosedChannelException ignored) {
        return -1;
    }
}

final ByteBuffer _internalNioBuffer(int index, int length, boolean duplicate) {
    index = idx(index);// offset+writerIndex
    ByteBuffer buffer = duplicate ? newInternalNioBuffer(memory) :
                                                        internalNioBuffer();
    buffer.limit(index + length).position(index);
    return buffer;
}
//通过 offset 获取使用内存的初始位置
```

```
protected final int idx(int index) {
    return offset + index;
}
//衍生一个ByteBuffer,与原ByteBuffer对象共享内存
protected ByteBuffer newInternalNioBuffer(ByteBuffer memory) {
    return memory.duplicate();
}
```

PoolByteBuf 通过 PoolSubpage 分配内存后返回的指针来获取偏移量,通过偏移量可以操作内存块的读/写,但还缺少对 PoolByteBuf 的初始化、指针 handle 如何转换成偏移量 offset、PoolByteBuf 与 PoolSubpage 的关联的介绍。

通过前面对 PoolChunk 的 allocateSubpage()方法进行的剖析,并结合本小节 PoolSubpage 的 allocateSubpage()方法的解读,对于小于 8KB 内存的第一次分配有了深入的了解,PoolSubpage 与 PoolByteBuf 的关联还需要回到 PoolChunk 的 allocate()方法。指针 handle 的高 32 位存储 PoolSubpage 中分配的内存段在 page 中的相对偏移量;低 32 位存储 page 在 PoolChunk 的二叉树中的位置 memoryMapIdx,通过这个位置获取其偏移量,两个偏移量相加就是 PoolByteBuf 的偏移量,PoolByteBuf 运用偏移量可操作读/写索引,实现数据的读/写。具体代码解读如下:

```
//内存具体分配方法
boolean allocate(PooledByteBuf<T> buf, int reqCapacity, int normCapacity) {
    final long handle;
    //内存指针,分配的二叉树内存节点偏移量或page和PoolSubpage的偏移量
    if ((normCapacity & subpageOverflowMask) != 0) { // 分配大于或等于page内存
        handle =  allocateRun(normCapacity);//具体分配的内存节点的偏移量
    } else {
        handle = allocateSubpage(normCapacity);// page 和 PoolSubpage 的偏移量
    }
    if (handle < 0) {//分配失败
        return false;
    }
    //从缓存的ByteBuffer对象池中获取一个ByteBuffer对象,有可能为null
    ByteBuffer nioBuffer = cachedNioBuffers != null ? cachedNioBuffers.pollLast() : null;
    //初始化申请到的内存数据,并对PoolByteBuf对象进行初始化
    initBuf(buf, nioBuffer, handle, reqCapacity);
```

```
        return true;
}
    void initBuf(PooledByteBuf<T> buf, ByteBuffer nioBuffer, long handle, int reqCapacity)
{
    //用int强转,取handle的低32位,低32位存储的是二叉树节点的位置
    int memoryMapIdx = memoryMapIdx(handle);
    //右移32位并强制转为int型,相当于获取handle的高32位,
    //即PoolSubpage的内存段相对page的偏移量
    int bitmapIdx = bitmapIdx(handle);
    if (bitmapIdx == 0) {//无PoolSubpage,即大于或等于page内存分配
        byte val = value(memoryMapIdx);//获取节点的高度
        assert val == unusable : String.valueOf(val);//判断节点高度是不可用(默认为12)
            //计算偏移量,offset的值为内存对齐偏移量
        buf.init(this, nioBuffer, handle, runOffset(memoryMapIdx) + offset,
            reqCapacity, runLength(memoryMapIdx), arena.parent.threadCache());
    } else {
        initBufWithSubpage(buf, nioBuffer, handle, bitmapIdx, reqCapacity);
    }
}
    //在6.2节讲过获取偏移量算法
private int runOffset(int id) {
    int shift = id ^ 1 << depth(id);
    return shift * runLength(id);
}
void initBufWithSubpage(PooledByteBuf<T> buf, ByteBuffer nioBuffer,
    long handle, int reqCapacity) {
    initBufWithSubpage(buf, nioBuffer, handle, bitmapIdx(handle), reqCapacity);
}
private void initBufWithSubpage(PooledByteBuf<T> buf, ByteBuffer nioBuffer,
                    long handle, int bitmapIdx, int reqCapacity) {
    assert bitmapIdx != 0;
    int memoryMapIdx = memoryMapIdx(handle);
    PoolSubpage<T> subpage = subpages[subpageIdx(memoryMapIdx)];
    assert subpage.doNotDestroy;
    assert reqCapacity <= subpage.elemSize;
    //根据page相对chunk的偏移量+PoolSubpage相对page的偏移量+对外内存地址对齐偏移量
    //初始化PoolByteBuf
```

```
buf.init(
    this, nioBuffer, handle,
    runOffset(memoryMapIdx) + (bitmapIdx & 0x3FFFFFFF) * subpage.elemSize +
        offset, reqCapacity, subpage.elemSize, arena.parent.threadCache());
}
```

内存的释放相对其分配来说要简单很多,下面主要剖析 PoolChunk 和 PoolSubpage 的释放。当内存释放时,同样先根据 handle 指针找到内存在 PoolChunk 和 PoolSubpage 中的相对偏移量,具体释放步骤如下。

(1)若在 PoolSubpage 上的偏移量大于 0,则交给 PoolSubpage 去释放,这与 PoolSubpage 内存申请有些相似,根据 PoolSubpage 内存分配段的偏移位 bitmapIdx 找到 long[]数组 bitmap 的索引 q,将 bitmap[q]的具体内存占用位 r 置为 0(表示释放)。同时调整 Arena 中的 PoolSubpage 缓存池,若 PoolSubpage 已全部释放了,且池中除了它还有其他节点,则从池中移除;若由于之前 PoolSubpage 的内存段全部分配完并从池中移除过,则在其当前可用内存段 numAvail 等于-1 且 PoolSubpage 释放后,对可用内存段进行"++"运算,从而使 numAvail++等于 0,此时会把释放的 PoolSubpage 追加到 Arena 的 PoolSubpage 缓存池中,方便下次直接从缓冲池中获取。

(2)若在 PoolSubpage 上的偏移量等于 0,或者 PoolSubpage 释放完后返回 false(PoolSubpage 已全部释放完,同时从 Arena 的 PoolSubpage 缓存池中移除了),则只需更新 PoolChunk 二叉树对应节点的高度值,并更新其所有父节点的高度值及可用字节数即可。

PoolChunk 的内存释放代码解读如下:

```
void free(long handle, ByteBuffer nioBuffer) {
int memoryMapIdx = memoryMapIdx(handle);
int bitmapIdx = bitmapIdx(handle);
if (bitmapIdx != 0) { // 先释放 subpage
    PoolSubpage<T> subpage = subpages[subpageIdx(memoryMapIdx)];
    assert subpage != null && subpage.doNotDestroy;
    //与分配时一样,先去 Arena 池中找到 subpage 对应的 head 指针
    PoolSubpage<T> head = arena.findSubpagePoolHead(subpage.elemSize);
```

```java
            synchronized (head) {
                //获取32位bitmapIdx交给PoolSubpage释放。释放后返回true,不再继续释放
                if (subpage.free(head, bitmapIdx & 0x3FFFFFFF)) {
                    return;
                }
            }
        }
        //释放的字节数调整
        freeBytes += runLength(memoryMapIdx);
        //设置节点值为节点初始化值,depth()方法使用的是byte[] depthMap,此字节初始化后就不再改变了
        setValue(memoryMapIdx, depth(memoryMapIdx));
        //更新父节点的高度值
        updateParentsFree(memoryMapIdx);
        //把nioBuffer放入缓存队列中,以便下次直接使用
        if (nioBuffer != null && cachedNioBuffers != null &&cachedNioBuffers.size() <
            DEFAULT_MAX_CACHED_BYTEBUFFERS_PER_CHUNK) {
            cachedNioBuffers.offer(nioBuffer);
        }
}

private void setValue(int id, byte val) {
    memoryMap[id] = val;
}

private void updateParentsFree(int id) {
    //节点的值
    int logChild = depth(id) + 1;
    while (id > 1) {
        int parentId = id >>> 1;
        byte val1 = value(id);
        //相邻节点
        byte val2 = value(id ^ 1);
        // 第一次循环时与id对应的高度值一样,后续都要减1,表示子节点的值
        logChild -= 1;
    //若两个子节点都可分配,则该节点变回自己所在层的depth,表示该节点也可完全被分配
        if (val1 == logChild && val2 == logChild) {
            setValue(parentId, (byte) (logChild - 1));
        } else {
            byte val = val1 < val2 ? val1 : val2;
```

```
            //设置父节点的高度值为两个子节点的最小值
        setValue(parentId, val);
    }
    id = parentId;
}
```

PoolSubpage 的内存释放代码解读如下:

```
boolean free(PoolSubpage<T> head, int bitmapIdx) {
if (elemSize == 0) {
        return true;
}
int q = bitmapIdx >>> 6;//由于 long 型是 64 位,因此除以 64 就是 long[] bitmap 的下标
int r = bitmapIdx & 63;//找到 bitmap[q]对应的位置
assert (bitmap[q] >>> r & 1) != 0;//判断当前位是否为已分配状态
bitmap[q] ^= 1L << r;//把 bitmap[q]的 r 位设为 0,表示未分配
//将该位置设置为下一个可用的位置,这也是在分配时会发生 nextAvail 大于 0 的情况
setNextAvail(bitmapIdx);
//若之前没有可分配的内存,从池中移除了,则将 PoolSubpage 继续添加到 Arena 的缓存池中,以便下回分配
if (numAvail ++ == 0) {
    addToPool(head);
    return true;
}
//若还有未被释放的内存,则直接返回
if (numAvail != maxNumElems) {
    return true;
} else {
    // 若内存全部被释放了,且池中没有其他的 PoolSubpage ,则不从池中移除,直接返回
    if (prev == next) {
        return true;
    }
    //若池中还有其他节点,且当前节点内存已全部被释放,则从池中移除
    //并返回 false,对其上的 page 也会进行相应的回收
    doNotDestroy = false;
    removeFromPool();
    return false;
    }
}
```

6.4 PoolArena 内存管理

在之前的内存分配的讲解中多次提到 PoolArena。那么 PoolArena 与 PoolChunk、PoolSubpage、PoolByteBuf 有什么关系呢？

Netty 的 I/O 线程在读取 Channel 数据时，需要内存分配器 PooledByteBufAllocator 来分配内存，而最终具体的分配工作是由 PoolArena 完成的，PoolArena 是内存的管理入口。由于 Netty 应用服务是多线程高并发系统，所以为了减少多线程同时分配同一块内存的竞争、提高内存的分配效率，在默认情况下会创建多个 PoolArena 并放入 PoolThreadLocalCache 中。PoolArena 究竟是如何管理 PoolChunk 和 PoolSubpage，并为 PoolByteBuf 分配内存的呢？下面先看看 PoolArena 的内部结构，如图 6-8 所示。

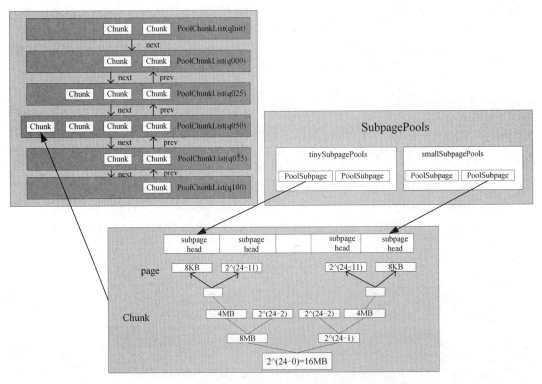

图 6-8 PoolArena 的内部结构

在图6-8中,除了左边的PoolChunkList,其他组件都已经详细的了解过了,tinySubpagePools 和 smallSubpagePools 分别缓存(0,512)和[512,8192)的 PoolSubpage 内存块。

PoolChunkList 是 PoolChunk 链表,内存在使用过程中会被重复利用。每次在分配内存时不会都去创建一个 PoolChunk,而是优先选择之前的 PoolChunk 内存块进行分配,若创建了多个 PoolChunk,则要考虑如何管理这些 PoolChunk,主要考虑内存能快速得到分配,而且不能过于浪费。PoolArena 根据 PoolChunk 的利用率,把 PoolChunk 划分到不同的 PoolChunkList 中。PoolChunkList 有两个属性,即 minUsage 和 maxUsage。当 PoolChunk 的利用率高于其所在的 PoolChunkList 的 maxUsage 的利用率时,PoolChunk 会从当前的 PoolChunkList 移动到下一个 PoolChunkList 中;否则会移动到上一个 PoolChunkList 中。这 6 个 PoolChunkList 的存储情况如下。

- qInit:存储内存利用率为[0%,25%)的 PoolChunk,qInit 的 preList 是其本身、nexList 为 q000;minUsage 为 Integer.MIN_VALUE。当 PoolChunk 在最开始创建时,如果其内存分配一直小于 25%,那么即使被完全释放,也不会被回收,会一直留在内存中。

- q000:存储内存利用率为[1%,50%)的 PoolChunk,其 preList 为 null、nexList 为 q025。PoolChunk 进入 q000 列表后,当其内存被完全释放,即内存利用率为 0 时,从 q000 链表中直接删除 PoolChunk,释放物理内存,主要是为了避免 PoolChunk 越来越多,导致内存被占满。

- q025:存储内存利用率为[25%,75%)的 PoolChunk,其 preList 为 q000、nexList 为 q050。为了避免 PoolChunk 在临界点时来回地在 q000 和 q025 链表间移动,它们两个链表的存储范围有一定的重叠。

- q050:存储内存利用率为[50%,100%)的 PoolChunk,其 preList 为 q025、nexList 为 q075。为了能提高内存分配的成功率,同时让 PoolChunk 的利用率保持在一个较高的水平,PoolArena 在分配内存时,选择从 q050 链表开始,尤其是在高峰期,由于请求量是平时的好几倍,创建的 PoolChunk 也是平时的好几倍,若先从 q000 开始,则在高峰期创建的大量 PoolChunk 的回收概率会大大降低，延缓了内存的回收进度，造成大量内存的浪费。

- q075：存储内存利用率为[75%,100%)的 PoolChunk，其 preList 为 q50、nexList 为 q100。它的内存利用率比较高，主要是在 q050 与 q100 间做个缓冲，为了让内存为利用率等于 100%的 PoolChunk 在回收了一小部分内存时不会很快进入 q050，否则下回分配可能又会被优选选中，导致内存利用率一直在 100%的边缘。因此 PoolArena 在分配内存时把 q075 放在了最后。
- q100：存储内存利用率为 100%的 PoolChunk，其 preList 为 q075、nexList 为 null，无法再继续分配，只能等待释放。

PoolArena 的内存分配关键源码解读如下：

```
private void allocate(PoolThreadCache cache, PooledByteBuf<T> buf, final int reqCapacity) {
//先把请求内存优化成内存池中的标准单元格大小，具体详细注解在后面
final int normCapacity = normalizeCapacity(reqCapacity);
if (isTinyOrSmall(normCapacity)) { //当小于 pageSize（默认为小于 8KB）时
    int tableIdx;
    PoolSubpage<T>[] table;
    boolean tiny = isTiny(normCapacity);//是否小于 512
    if (tiny) { // < 512
        //先尝试从线程本地缓存中获取
        if (cache.allocateTiny(this, buf, reqCapacity, normCapacity)) {
            // was able to allocate out of the cache so move on
            return;
        }
        //通过空间大小获取 tinySubpagePools 的下标
        //由于 tinySubpagePools 存储的是 16 的倍数的 PoolSubpage
        //因此 normCapacity/16=tableIndx
        tableIdx = tinyIdx(normCapacity);
        table = tinySubpagePools;
    } else {//大于或等于 512 且小于 8192
        if (cache.allocateSmall(this, buf, reqCapacity, normCapacity)) {
            // was able to allocate out of the cache so move on
```

```java
            return;
        }
        //通过空间大小获取smallSubpagePools的下标
        tableIdx = smallIdx(normCapacity);
        table = smallSubpagePools;
    }

    final PoolSubpage<T> head = table[tableIdx];
    synchronized (head) {//对head头部指针加锁
        final PoolSubpage<T> s = head.next;
        if (s != head) {
        //当头部指针与其next不同时，表示此PoolSubpages缓存中有内存可分配
            assert s.doNotDestroy && s.elemSize == normCapacity;
            //通过PoolSubpage分配获取偏移量信息
            long handle = s.allocate();
            assert handle >= 0;//判断是否分配成功
            //初始化PoolByteBuf
            s.chunk.initBufWithSubpage(buf, null, handle, reqCapacity);
            //增加对应分配的次数
            incTinySmallAllocation(tiny);
            return;
        }
    }
    synchronized (this) {//为此PoolArena加锁
        //若线程本地缓存和PoolSubpages中都没有可分配的内存
        //此分配方法详细注释在后面
        allocateNormal(buf, reqCapacity, normCapacity);
    }

    incTinySmallAllocation(tiny);
    return;
}
if (normCapacity <= chunkSize) {
    //尝试从线程本地缓存中获取
    if (cache.allocateNormal(this, buf, reqCapacity, normCapacity)) {
        return;
    }
    synchronized (this) {
        allocateNormal(buf, reqCapacity, normCapacity);
```

```
            ++allocationsNormal;
        }
    } else {
        //大内存分配时不放入缓存池
        allocateHuge(buf, reqCapacity);
    }
}
//内存被划分成固定大小的内存单元,会根据请求的内存进行计算匹配最接近的内存单元
int normalizeCapacity(int reqCapacity) {
    checkPositiveOrZero(reqCapacity, "reqCapacity");
    if (reqCapacity >= chunkSize) {//大于16MB
        return reqCapacity;
    }
    //大于512
    if (!isTiny(reqCapacity)) { // >= 512
        // Doubled
        int normalizedCapacity = reqCapacity;
        //防止reqCapacity为512、1024、2048等临界点翻倍,先进行减1操作
        //例如,当reqCapacity为512时,先减1变成511,再去寻找与其最接近的2的幂数
        //下面位移或操作主要是让其所有位都变成1
        //当reqCapacity为(512,1023]时,经过下面的位移或操作后,其拥有的所有01位都变成1
        // 即变成1023,最后再加1就成了1024
        //由于reqCapacity为整数,最多32位,因此此处的右移为(1,1*2,2*2,2*2*2,2*2*2*2)
        normalizedCapacity --;
        normalizedCapacity |= normalizedCapacity >>>  1;
        normalizedCapacity |= normalizedCapacity >>>  2;
        normalizedCapacity |= normalizedCapacity >>>  4;
        normalizedCapacity |= normalizedCapacity >>>  8;
        normalizedCapacity |= normalizedCapacity >>> 16;
        normalizedCapacity ++;
        //当溢出时会变成负数,此时需要右移1位
        if (normalizedCapacity < 0) {
            normalizedCapacity >>>= 1;
        }
        return normalizedCapacity;
    }
    // 当小于512且是16的整数倍时,直接返回
```

```java
    if ((reqCapacity & 15) == 0) {
        return reqCapacity;
    }
    //当小于12且不是16的整数倍时，低4位变成16
    return (reqCapacity & ~15) + 16;
}
//内存分配
private void allocateNormal(PooledByteBuf<T> buf, int reqCapacity,
    int normCapacity) {
//先从q050链表开始分配内存，从链表中循环取出PoolChunk，如果分配成功了，则返回true；否则返回false
    if (q050.allocate(buf, reqCapacity, normCapacity) ||
        q025.allocate(buf, reqCapacity,
            normCapacity) ||q000.allocate(buf, reqCapacity, normCapacity)||
            qInit.allocate(buf, reqCapacity, normCapacity) ||
        q075.allocate(buf, reqCapacity, normCapacity)) {
        return;
    }
    //5个链表都未能成功分配到内存，此时需要开辟一块新的PoolChunk内存
    PoolChunk<T> c = newChunk(pageSize, maxOrder, pageShifts, chunkSize);
        //此处为PoolChunk的内存分配，在前面小节中详细讲解过
    boolean success = c.allocate(buf, reqCapacity, normCapacity);
    assert success;
        //分配成功后，把PoolChunk追加到qInit链表中
    qInit.add(c);
}
```

PoolArena 除了内存分配，还管理内存的释放，内存释放代码如下：

```java
void free(PoolChunk<T> chunk, ByteBuffer nioBuffer, long handle, int
 normCapacity, PoolThreadCache cache) {
//非内存池内存释放比较简单，直接物理释放即可
if (chunk.unpooled) {
    int size = chunk.chunkSize();
    destroyChunk(chunk);
    activeBytesHuge.add(-size);
    deallocationsHuge.increment();
} else {
    SizeClass sizeClass = sizeClass(normCapacity);
```

```java
        //先尝试放入线程本地缓存
        //在线程本地缓存默认的情况下，缓存 tiny 类型的 PoolSubpage 数最多为 512
        //small 类型的 PoolSubpage 数最多为 256, normal 类型的 PoolSubpage 数最多为 64 个
        //这些配置都在 PooledByteBufAllocator 类中
        if (cache != null && cache.add(this, chunk, nioBuffer, handle, normCapacity,
            sizeClass)) {
            // cached so not free it.
            return;
        }
        //当未成功放入缓存时，释放 PoolChunk
        freeChunk(chunk, handle, sizeClass, nioBuffer, false);
    }
}
void freeChunk(PoolChunk<T> chunk, long handle, SizeClass sizeClass,
               ByteBuffer nioBuffer, boolean finalizer) {
    final boolean destroyChunk;
    synchronized (this) {//内存分配和回收都要在 Arena 上加锁
        if (!finalizer) {
            switch (sizeClass) {
                case Normal:
                    ++deallocationsNormal;
                    break;
                case Small:
                    ++deallocationsSmall;
                    break;
                case Tiny:
                    ++deallocationsTiny;
                    break;
                default:
                    throw new Error();
            }
        }
        //chunk.parent 为 chunk 所在的 PoolChunkList
        //调用 PoolChunkList 的 free() 方法回收内存块，若内存完全空闲，则销毁
        destroyChunk = !chunk.parent.free(chunk, handle, nioBuffer);
    }
    if (destroyChunk) {
```

```
        //当 finalizer 为 true, 或者 PoolChunkList 为 q000 时,释放后的内存利用率为 0
            //此时需要物理释放
        destroyChunk(chunk);
    }
}

    //物理释放
    @Override
protected void destroyChunk(PoolChunk<ByteBuffer> chunk) {
    if (PlatformDependent.useDirectBufferNoCleaner()) {
        PlatformDependent.freeDirectNoCleaner(chunk.memory);
    } else {
        PlatformDependent.freeDirectBuffer(chunk.memory);
    }
}

    //PoolChunkList 的 free()方法
    boolean free(PoolChunk<T> chunk, long handle, ByteBuffer nioBuffer) {
    //先调用 chunk 的 free()方法,把内存标记为已释放
    chunk.free(handle, nioBuffer);
    if (chunk.usage() < minUsage) {
        //若内存利用率小于 minUsage
        //则此时需要把 PoolChunk 从当前 PoolChunkList 中移除
        remove(chunk);
        //把移除的 PoolChunk 移动到前一个 PoolChunkList 中
        return move0(chunk);
    }
    return true;
}

    //从 PoolChunk 链表中移除
    private void remove(PoolChunk<T> cur) {
    if (cur == head) {//若当前 chunk 为 PoolChunkList 的第一个
        head = cur.next;//把 chunk 的下一个元素变成 PoolChunkList 的第一个元素
        if (head != null) {
            head.prev = null;
        }
    } else {//修改指针指向
```

```
            PoolChunk<T> next = cur.next;
            cur.prev.next = next;//把chunk前面的chunk的next指针指向当前chunk的next
            if (next != null) { //若当前chunk的下一个chunk不为空
               //则把当前chunk的下一个chunk的prev指针从当前chunk改成当前chunk的前一个元素
                next.prev = cur.prev;
            }
        }
    }
    private boolean move0(PoolChunk<T> chunk) {
        //在当前PoolChunkList为q000时,直接物理释放
        if (prevList == null) {
            assert chunk.usage() == 0;
            return false;
        }
        //把PoolChunk移到前面的PoolChunkList中
        return prevList.move(chunk);
    }
```

除了线程本地缓存 PoolThreadCache,关于 PoolArena 的内存分配和回收基本上已经了解了。PoolThreadCache 中有多个 MemoryRegionCache 数组,每种类型的内存都有一个 MemoryRegionCache 数组与之对应。MemoryRegionCache 中有个队列,这个队列主要是用来存放内存对象的,具体源码此处就不讲解了。

整体内存分配时序图如图 6-9 所示,在应用 Netty 时,通过默认设置 PooledByteBufAllocator 执行 ByteBuf 的分配。当用 NioByteUnsafe 的 read()方法读取 NioSocketChannel 数据时,需要调用 PooledByteBufAllocator 去分配内存,具体分配多少内存,由 Handle 的 guess()方法决定,此方法只预测所需的缓冲区的大小,不进行实际的分配。PooledByteBufAllocator 从 PoolThreadLocalCache 中获取 PoolArena,最终的内存分配工作由 PoolArena 完成。

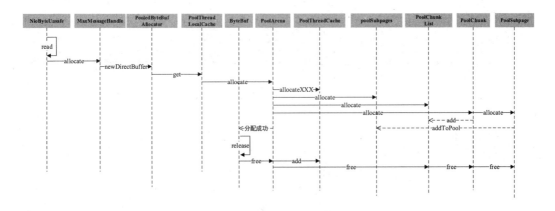

图 6-9　整体内存分配时序图

6.5　RecvByteBufAllocator 内存分配计算

虽然了解了 Netty 整个内存池管理的细节（包括它的内存分配的具体逻辑），但是每次在从 NioSocketChannel 中读取数据时，应该分配多少内存去读呢？例如，客户端发送的数据为 1KB，若每次都分配 8KB 的内存去读取数据，则会导致内存大量浪费；若分配 16B 的内存去读取数据，那么需要 64 次才能全部读完，对性能有很大的影响。那么对于这个问题，Netty 是如何解决的呢？

NioEventLoop 线程在处理 OP_READ 事件，进入 NioByteUnsafe 循环读取数据时，使用了两个类来处理内存的分配：一个是 ByteBufAllocator，PooledByteBufAllocator 为它的默认实现类；另一个是 RecvByteBufAllocator，AdaptiveRecvByteBufAllocator 是它的默认实现类，在 DefaultChannelConfig 初始化时设置。PooledByteBufAllocator 主要用来处理内存的分配，并最终委托 PoolArena 去完成。AdaptiveRecvByteBufAllocator 主要用来计算每次读循环时应该分配多少内存。NioByteUnsafe 之所以需要循环读取，主要是因为分配的初始 ByteBuf 不一定能够容纳读取到的所有数据。NioByteUnsafe 循环读取的核心代码解读如下：

```
    @Override
public final void read() {
    final ChannelConfig config = config();
```

```
    ….
    //获取内存分配器,默认是 PooledByteBufAllocator
final ByteBufAllocator allocator = config.getAllocator();
    //获取 RecvByteBufAllocator 内部的计算器 Handle
final RecvByteBufAllocator.Handle allocHandle = recvBufAllocHandle();
allocHandle.reset(config);
    do {
            // allocator 根据计算器 Handle 计算此次需要分配多少内存并从内存池中分配
        byteBuf = allocHandle.allocate(allocator);
            //设置最后一次分配的内存大小加上每次读取的字节数
        allocHandle.lastBytesRead(doReadBytes(byteBuf));
                pipeline.fireChannelRead(byteBuf);
        …
    } while (allocHandle.continueReading());
    //读结束后调用,记录此次实际读取到的数据的大小,并预测下一次内存分配的大小
    allocHandle.readComplete();
}
```

RecvByteBufAllocator 的默认实现类 AdaptiveRecvByteBufAllocator 是实际的缓冲管理区,这个类可以根据读取到的数据预测所需字节的多少,从而自动增加或减少;如果上一次读循环将缓冲区填充满了,那么预测的字节数会变大。如果连续两次读循环都不能填满已分配的缓冲区,则预测的字节数会变小。

AdaptiveRecvByteBufAllocator 内部维护了一个 SIZE_TABLE 数组,记录了不同的内存块大小,按照分配需要寻找最合适的内存块。SIZE_TABLE 数组中的值都是 2^n,这样便于软硬件进行处理,SIZE_TABLE 数组的初始化与 PoolArena 中的 normalizeCapacity 的初识化类似。当需要的内存很小时,增长的幅度不大;当需要的内存较大时,增长幅度比较大。因此在[16,512]区间每次增加 16,直到 512;而从 512 起,每次翻一倍,直到 int 的最大值。

当对内部计算器 Handle 的具体实现类 HandleImpl 进行初始化时,可根据 AdaptiveRecvByteBufAllocator 的 getSizeTableIndex 二分查找方法获取 SIZE_TABLE 的下标 index 并保存,通过 SIZE_TABLE [index] 获取下次需要分配的缓冲区的大小 nextReceiveBufferSize 并记录。缓冲区的最小容量属性对应 SIZE_TABLE 中的下标为 minIndex 的值,最大容量属性对应 SIZE_TABLE 中

的下标为 maxIndex 的值及 bool 类型标识属性 decreaseNow。这 3 个属性用于判断下一次创建的缓冲区是否需要减小。

NioByteUnsafe 每次读循环完成后会根据实际读取到的字节数和当前缓冲区的大小重新设置下次需要分配的缓冲区的大小，具体代码解读如下：

```
    //循环读取完后被调用
    @Override
public void readComplete() {
    record(totalBytesRead());
}
    private void record(int actualReadBytes) {
    if (actualReadBytes <= SIZE_TABLE[max(0, index - INDEX_DECREMENT - 1)]) {
        if (decreaseNow) {//若减小标识 decreaseNow 连续两次为 true，则说明下次读取字节数需要减小
            // SIZE_TABLE 下标减 1
            index = max(index - INDEX_DECREMENT, minIndex);
            nextReceiveBufferSize = SIZE_TABLE[index];
            decreaseNow = false;
        } else {
            //第一次减小，只做记录
            decreaseNow = true;
        }
        //实际读取的字节大小要大于或等于预测值
      } else if (actualReadBytes >= nextReceiveBufferSize) {
    // SIZE_TABLE 下标加 4
        index = min(index + INDEX_INCREMENT, maxIndex);
        //若当前缓存为 512，则变成 512*2^4
    nextReceiveBufferSize = SIZE_TABLE[index];
    decreaseNow = false;
    }
}
```

可以模拟 NioByteUnsafe 的 read() 方法，在每次读循环开始时，一定要先重置 totalMessages 与 totalBytesRead（清零），读取完成后，readComplete 会计算并调整下次预计需要分配的缓冲区的大小，具体代码如下：

```java
@Test
public void guessTest(){
AdaptiveRecvByteBufAllocator alloctor = new AdaptiveRecvByteBufAllocator();
RecvByteBufAllocator.Handle handle = alloctor.newHandle();
System.out.println("========开始I/O读事件模拟=============");
//读取循环开始前先重置，将读取的次数和字节数设置为0
//将totalMessages与totalBytesRead置0
handle.reset(null);
System.out.println(String.format("第1次模拟读,需要分配的大小：%d",
 handle.guess()));
handle.lastBytesRead(256);
//调整下次预测值
handle.readComplete();
//在每次读数据时都需要重置totalMessages与totalBytesRead
handle.reset(null);
System.out.println(String.format("第2次模拟读,需要分配的大小：%d",
 handle.guess()));
handle.lastBytesRead(256);
handle.readComplete();
System.out.println("======连续2次读取的字节数小于默认分配的字节数=====");
handle.reset(null);
System.out.println(String.format("第3次模拟读,需要分配的大小：%d",
 handle.guess()));
handle.lastBytesRead(512);
//调整下次预测值,预测值应该增加到512*2^4
handle.readComplete();
System.out.println("=====读取的字节数变大=====");
handle.reset(null);
  // 读循环中缓冲区变大
System.out.println(String.format("第4次模拟读,需要分配的大小：%d",
 handle.guess()));
}
```

6.6 小结

　　内存管理是 Netty 最核心的部分，但也是最难以理解的部分，这主要是因为代码中涉及多线程的并发，以及各种位移的与或非操作。建议读者在学习时，把难以理解的代码贴到测试类中进行简单的调试。本章首先介绍了底层 PoolChunk 的内存分配，然后介绍了上层的 PoolArena 对内存的整体管理，最后介绍了 NioEventLoop 处理 NioSocketChannel 的 OP_READ 事件过程。当进行循环读数据时，运用 ByteBufAllocator（内存分配器）与 AdaptiveRecvByteBufAllocator 计算读循环需要分配的内存，把整个内存分配与应用结合了起来。至此，Netty 的内存管理基本上介绍完了。

第 7 章

Netty 时间轮高级应用

Netty 主要应用于网络通信，本书完成了一套分布式 RPC 框架，实现了服务之间的长连接通信。Netty 还有一个非常重要的应用领域——即时通信系统 IM。在 IM 聊天系统中，有成千上万甚至百万条的链路，Netty 是如何管理这些链路的呢？Netty 有一套自带的心跳检测机制，这套检测机制的原理是通过创建多个定时任务 ScheduledFutureTask，定时一段时间与客户端进行通信，确保连接可用。

除了自带的心跳检测机制，Netty 还提供了另外一种方案，叫时间轮 HashedWheelTimer。时间轮也是一个定时任务，只是这个定时任务是额外的一条线程，每隔一段时间运行一次，在运行过程中，只会把当前时间要执行的任务捞出来运行，而并不会去捞那些还未到时的任务。

7.1　Netty 时间轮的解读

都知道时钟有指针、刻度、每刻度表示的时长等属性，Netty 时间轮的设计也差不多，只是时钟的指针有时、分、秒，而 Netty 只用了一个指针。那么 Netty 是如何把定时任务加入时间轮的呢？下面先看一幅时间轮的构造图，如图 7-1 所示。

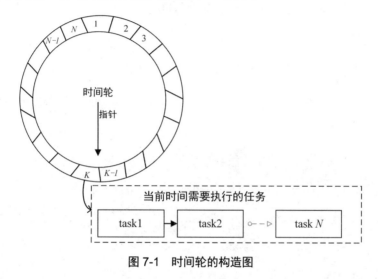

图 7-1　时间轮的构造图

从图 7-1 中可以看出，当指针指向某一刻度时，它会把此刻度中的所有 task 任务一一取出并运行。在解读 Netty 的时间轮代码前，先思考以下 3 个问题。

- 时间轮的指针走一轮是多久？
- 时间轮是采用什么容器存储这些 task 的？
- 定时任务的运行时间若晚于指针走一轮的终点（见图 7-1 中的 N），则此时此任务该放在哪个刻度？

（1）刻度的间隔时间标注为 tickDuration，同时将时间轮一轮的刻度总数标注为 wheelLen，两者都是时间轮的属性，可以通过构造方法由使用者传入，这样就可以得到时间轮指针走一轮的时长=tickDuration*wheelLen。

（2）当指针运行到某一刻度时，需要把映射在此刻度上所有的任务都取出来，而刻度总数在时间轮初始化后就固定了。因此与 Map 相似，采用数组标识 wheel[]加链表的方式来存储这些 task，数组的大小固定为图 7-1 中的 N，刻度的编号就是 wheel[]的下标。

（3）每个时间轮启动都会记录其启动时间，同时，每个定时任务都有其确定的执行时间，用这个执行时间减去时间轮的启动时间，再除以刻度的持续时长，就能获取这个定时任务需要指针走过多少刻度才运行，标注为 calculated。

时间轮本身记录了当前指针已经走过了多少刻度，标注为 tick。通过 calculated、tick、时间轮刻度总数 wheelLen 来计算定时任务在哪一刻度执行（此刻度标注为 stopIndex）。需要分为以下几种情况进行处理。

- 当 calculated<tick 时，说明这项任务已经是旧任务了，可立刻执行，因此 stopIndex=tick。
- 当(calculated-tick)<=wheelLen 时，stopIndex=(calculated-tick)。
- 当(calculated-tick)>wheelLen 时，calculated 肯定大于 wheelLen，若 wheelLen 是 2 的整数次幂，则可以运用与运算 stopIndex=calculated& (wheelLen-1)；若 wheelLen 不是 2 的整数次幂，则把它转换成距离最近的 2 的整数次幂即可。

7.1.1 时间轮源码剖析之初始化构建

经过对以上 3 个问题进行的分析，对时间轮的构造有了基本的认知，了解了时间轮内部属性特征，以及定时任务与刻度的映射关系。但具体时间轮是如何运行的，它的指针是如何跳动的。这都需要通过仔细阅读 Netty 的时间轮源码来寻找答案。时间轮源码分为两部分：第一部分包含时间轮的核心属性、初始化构建、启动和定时检测任务的添加；第二部分主要是对时间轮的时钟 Worker 线程的剖析。线程的核心功能有时钟指针的刻度跳动、超时任务处理、任务的取消等。

时间轮 HashedWheelTimer 的核心属性解读如下：

```
//时间轮实例个数
private static final AtomicInteger INSTANCE_COUNTER = new AtomicInteger();
```

```java
//在服务过程中,时间轮实例个数不能超过 64 个
private static final AtomicBoolean WARNED_TOO_MANY_INSTANCES = new
                                                    AtomicBoolean();
private static final int INSTANCE_COUNT_LIMIT = 64;
//刻度持续时长最小值,不能小于这个最小值
private static final long MILLISECOND_NANOS =
                                        TimeUnit.MILLISECONDS.toNanos(1);
//内存泄漏检测
private static final ResourceLeakDetector<HashedWheelTimer> leakDetector =
                ResourceLeakDetectorFactory.instance()
                        .newResourceLeakDetector(HashedWheelTimer.class, 1);
//原子性更新时间轮工作状态,防止多线程重复操作
private static final AtomicIntegerFieldUpdater<HashedWheelTimer>
        WORKER_STATE_UPDATER =
        AtomicIntegerFieldUpdater.newUpdater(HashedWheelTimer.class,
                                                        "workerState");
//内存泄漏检测虚引用
private final ResourceLeakTracker<HashedWheelTimer> leak;
//用于构建时间轮工作线程的 Runnable 掌控指针的跳动
private final HashedWheelTimer.Worker worker =
                                        new HashedWheelTimer.Worker();
//时间轮工作线程
private final Thread workerThread;
//时间轮的 3 种工作状态分别为初始化、已经启动正在运行、停止
public static final int WORKER_STATE_INIT = 0;
public static final int WORKER_STATE_STARTED = 1;
public static final int WORKER_STATE_SHUTDOWN = 2;
private volatile int workerState;
//每刻度的持续时间
private final long tickDuration;
//此数组用于存储映射在时间轮刻度上的任务
private final HashedWheelTimer.HashedWheelBucket[] wheel;
//时间轮总格子数-1
private final int mask;
//同步计数器,时间轮 Worker 线程启动后,将同步给调用时间轮的线程
private final CountDownLatch startTimeInitialized = new CountDownLatch(1);
/**
```

```
 * 超时task任务队列,先将任务放入这个队列中
 * 再在Worker线程中从队列中取出并放入wheel[]的链表中
 */
private final Queue<HashedWheelTimer.HashedWheelTimeout> timeouts =
                                            PlatformDependent.newMpscQueue();
    /**
     * 取消的task任务存放队列
     * 在Worker线程中会检测是否有任务需要取消
     * 若有,则找到对应的链表,
     * 并修改这些取消任务的前后任务的指针
     */
private final Queue<HashedWheelTimer.HashedWheelTimeout> cancelledTimeouts =
                                    PlatformDependent.newMpscQueue();
//目前需等待执行的任务数
private final AtomicLong pendingTimeouts = new AtomicLong(0);
//时间轮最多容纳多少定时检测任务,默认为-1,无限制
private final long maxPendingTimeouts;
//时间轮启动时间
private volatile long startTime;
```

时间轮 HashedWheelTimer 的核心构造函数解读如下:

```
    /**
     * 时间轮构造函数
     * @param threadFactory 线程工厂,用于创建线程
     * @param tickDuration 刻度持续时长
     * @param unit    刻度持续时长单位
     * @param ticksPerWheel 时间轮总刻度数
     * @param leakDetection 是否开启内存泄漏检测
     * @param maxPendingTimeouts 时间轮可接受最大定时检测任务数
     */
    public HashedWheelTimer(
        ThreadFactory threadFactory,
        long tickDuration, TimeUnit unit, int ticksPerWheel,
        boolean leakDetection,
        long maxPendingTimeouts) {
      /**
       * 对时间轮刻度数进行格式化,转换成离ticksPerWheel最近的2的整数次幂
```

```java
 * 并初始化 wheel 数组
 */
wheel = createWheel(ticksPerWheel);
mask = wheel.length - 1;
//把刻度持续时长转换成纳秒，这样更加精确
long duration = unit.toNanos(tickDuration);
/**
 * 检测持续时长不能太长，但也不能太短
 */
if (duration >= Long.MAX_VALUE / wheel.length) {
    throw new IllegalArgumentException(String.format(
            "tickDuration: %d (expected: 0 < tickDuration in nanos < %d",
            tickDuration, Long.MAX_VALUE / wheel.length));
}
if (duration < MILLISECOND_NANOS) {
    if (logger.isWarnEnabled()) {
        logger.warn("Configured tickDuration %d smaller then %d," +
                    "using 1ms.", tickDuration, MILLISECOND_NANOS);
    }
    this.tickDuration = MILLISECOND_NANOS;
} else {
    this.tickDuration = duration;
}
//构建时间轮的 Worker 线程
workerThread = threadFactory.newThread(worker);
//是否需要内存泄漏检测
leak = leakDetection || !workerThread.isDaemon() ?
                    leakDetector.track(this) : null;
//最大定时检测任务个数
this.maxPendingTimeouts = maxPendingTimeouts;
//时间轮实例个数检测，超过 64 个会告警
if (INSTANCE_COUNTER.incrementAndGet() > INSTANCE_COUNT_LIMIT &&
        WARNED_TOO_MANY_INSTANCES.compareAndSet(false, true)) {
    reportTooManyInstances();
}
}
/**
```

```
 * 格式化总刻度数
 * 初始化时间轮容器
 * @param ticksPerWheel
 * @return
 */
private static HashedWheelTimer.HashedWheelBucket[] createWheel(
int ticksPerWheel) {
    //格式化
    ticksPerWheel = normalizeTicksPerWheel(ticksPerWheel);
    HashedWheelTimer.HashedWheelBucket[] wheel = new
                HashedWheelTimer.HashedWheelBucket[ticksPerWheel];
    for (int i = 0; i < wheel.length; i ++) {
        wheel[i] = new HashedWheelTimer.HashedWheelBucket();
    }
    return wheel;
}
/**
 * 找到离 ticksPerWheel 最近的 2 的整数次幂
 * @param ticksPerWheel
 * @return
 */
private static int normalizeTicksPerWheel(int ticksPerWheel) {
    int normalizedTicksPerWheel = 1;
    while (normalizedTicksPerWheel < ticksPerWheel) {
        normalizedTicksPerWheel <<= 1;
    }
    return normalizedTicksPerWheel;
}
```

时间轮 HashedWheelTimer 添加定时任务及其启动代码解读如下：

```
public Timeout newTimeout(TimerTask task, long delay, TimeUnit unit) {
    /**
     * 需等待执行的任务数+1
     * 同时判断是否超过了最大限制
     */
    long pendingTimeoutsCount = pendingTimeouts.incrementAndGet();
    if (maxPendingTimeouts > 0 && pendingTimeoutsCount > maxPendingTimeouts) {
```

```java
        pendingTimeouts.decrementAndGet();
        throw new RejectedExecutionException();
    }
    //若时间轮Worker线程未启动,则需启动
    start();
    /**
     * 根据定时任务延时执行时间与时间轮启动时间
     * 获取相对时间轮开始后的任务执行延时时间
     * 因为时间轮开始启动时间是不会改变的,所以通过这个时间可获取时钟需要跳动的刻度
     */
    long deadline = System.nanoTime() + unit.toNanos(delay) - startTime;
    // Guard against overflow.
    if (delay > 0 && deadline < 0) {
        deadline = Long.MAX_VALUE;
    }
    /**
     * 构建定时检测任务,并将其添加到新增定时检测任务队列中
     * 在Worker线程中,会从队列中取出定时检测任务并放入缓存数组wheel中
     */
    HashedWheelTimer.HashedWheelTimeout timeout =
            new HashedWheelTimer.HashedWheelTimeout(this, task, deadline);
    timeouts.add(timeout);
    return timeout;
}

/**
 * 时间轮启动
 */
public void start() {
    //根据时间轮的状态进行对应的处理
    switch (WORKER_STATE_UPDATER.get(this)) {
        //当时间轮处于初始化状态时,需要启动它
        case WORKER_STATE_INIT:
            //原子性启动
            if (WORKER_STATE_UPDATER.compareAndSet(this, WORKER_STATE_INIT,
                WORKER_STATE_STARTED)) {
                workerThread.start();
```

```
        }
        break;
    case WORKER_STATE_STARTED:
        break;
    case WORKER_STATE_SHUTDOWN:
        throw new IllegalStateException("cannot be started once stopped");
    default:
        throw new Error("Invalid WorkerState");
    }
    // 等待 Worker 线程初始化成功
    while (startTime == 0) {
        try {
            startTimeInitialized.await();
        } catch (InterruptedException ignore) {
        }
    }
}
```

7.1.2　时间轮源码剖析之 Worker 启动线程

Worker 线程是整个时间轮的核心，它拥有一个属性——tick。tick 与时间刻度有一定的关联，指针每经过一个刻度后，tick++；tick 与 mask（时间轮总格子数-1）进行与操作后，就是时间轮指针的当前刻度序号。在 Worker 线程中，tick 做了以下 4 件事。

- 等待下一刻度运行时间到来。
- 从取消任务队列中获取需要取消的任务并处理。
- 从任务队列中获取需要执行的定时检测任务，并把它们放入对应的刻度链表中。
- 从当前刻度链表中取出需要执行的定时检测任务，并循环执行这些定时检测任务的 run() 方法。

具体代码解读如下：

```
//当调用了时间轮的 stop()方法后,将获取其未执行完的任务
private final Set<Timeout> unprocessedTimeouts = new HashSet<Timeout>();
//时钟指针跳动次数
```

```java
private long tick;
@Override
public void run() {
    //时间轮启动时间
    startTime = System.nanoTime();
    if (startTime == 0) {
        startTime = 1;
    }
    //Worker 线程初始化了，通知调用时间轮启动的线程
    startTimeInitialized.countDown();
    do {
        /**
         * 获取下一刻度度时间轮总体的执行时间
         * 当这个时间与时间轮启动时间的和大于当前时间时，线程会睡眠到这个时间点
         */
        final long deadline = waitForNextTick();
        if (deadline > 0) {
            //获取刻度的编号，即 wheel 数组的下标
            int idx = (int) (tick & mask);
            //先处理需要取消的任务
            processCancelledTasks();
            //获取刻度所在的缓存链表
            HashedWheelTimer.HashedWheelBucket bucket =
                    wheel[idx];
            //把新增的定时检测任务加入 wheel 数组的缓存链表中
            transferTimeoutsToBuckets();
            //循环执行刻度所在的缓存链表
            bucket.expireTimeouts(deadline);
            //执行完后，指针才正式跳动
            tick++;
        }
        //时间轮状态需要为已启动状态
    } while (WORKER_STATE_UPDATER.get(HashedWheelTimer.this) ==
                        WORKER_STATE_STARTED);
    //运行到这里说明时间轮停止了，需要把未处理的任务返回
    for (HashedWheelTimer.HashedWheelBucket bucket: wheel) {
        bucket.clearTimeouts(unprocessedTimeouts);
```

```java
        }
        /**
         * 刚刚加入还未来得及放入时间轮缓存中的超时任务
         * 也需要捞出并放入 unprocessedTimeouts 中一起返回
         */
        for (;;) {
            HashedWheelTimer.HashedWheelTimeout timeout = timeouts.poll();
            if (timeout == null) {
                break;
            }
            if (!timeout.isCancelled()) {
                unprocessedTimeouts.add(timeout);
            }
        }
        //处理需要取消的任务
        processCancelledTasks();
}
private long waitForNextTick() {
        //获取下一刻度时间轮总体的执行时间
        long deadline = tickDuration * (tick + 1);
        for (;;) {
            //当前时间-启动时间
            final long currentTime = System.nanoTime() - startTime;
            /**
             * 计算需要睡眠的毫秒时间
             * 由于在将纳秒转换为毫秒时需要除以 1000000
             * 因此需要加上 999999，以防丢失尾数，任务被提前执行
             */
            long sleepTimeMs = (deadline - currentTime + 999999) / 1000000;
            /**
             * 当睡眠时间小于 0 且
             * 等于 Long.MIN_VALUE 时，直接跳过此刻度
             * 否则不睡眠，直接执行任务
             */
            if (sleepTimeMs <= 0) {
                if (currentTime == Long.MIN_VALUE) {
                    return -Long.MAX_VALUE;
```

```
        } else {
            return currentTime;
        }
    }
    /**
     * Windows 操作系统特殊处理
     * 其 Sleep 函数是以 10ms 为单位进行延时的
     * 也就是说,所有小于 10 且大于 0 的情况都是 10ms
     * 所有大于 10 且小于 20 的情况都是 20ms
     * 因此这里做了特殊处理,对于小于 10ms 的,直接不睡眠;对于大于 10ms 的,去掉尾数
     */
    if (PlatformDependent.isWindows()) {
        sleepTimeMs = sleepTimeMs / 10 * 10;
    }
    try {
        Thread.sleep(sleepTimeMs);
    } catch (InterruptedException ignored) {
        //当发生异常,发现时间轮状态为 WORKER_STATE_SHUTDOWN 时,立刻返回
        if (WORKER_STATE_UPDATER.get(HashedWheelTimer.this) ==
          WORKER_STATE_SHUTDOWN) {
            return Long.MIN_VALUE;
        }
    }
}
```

至于其他方法,包括处理取消任务 processCancelledTasks(),以及从超时任务队列中获取任务并放入时间轮缓存链表中 transferTimeoutsToBuckets() 等方法的源码解读,比较容易看懂,此处不再赘述。

7.2　Netty 时间轮改造方案制订

通过 7.1 节的学习,虽然没有实际的运用时间轮,但是对它有了比较深入的了解。采用时间轮进行心跳检测的实现思路大概如下。

每条 I/O 线程都会构建一个时间轮,当然也可以只构建一个静态的时间轮,根据链路数量来决定。

当有 Channel 通道进来时,会触发 channelRegistered()方法,在此方法中,把通道的心跳定时检测任务交给时间轮,再调用其 newTimeout()方法把任务添加到时间轮中。

采用时间轮去执行这些定时任务,很明显可以减轻 I/O 线程的负担,但这些定时任务同样是放在内存中的,因此设置定时检测时间一定要注意不宜过长。虽然单机的长连接并发量不会太高,放在内存也不会有太大的影响。但是若除心跳检测外,用时间轮作为公司的任务定时调度系统或监控 10 亿级定时检测任务系统,则此时再放内存,恐怕再大的内存也会被撑爆,本节通过改造时间轮来解决这个问题?

以监控 10 亿级定时检测任务系统为例,想要实现这套系统,用传统轮询方式性能肯定无法满足要求;用时间轮来实现,若将每天 10 亿级任务存放到内存中,则肯定会发生内存溢出,但可以通过改造把时间轮的任务数据存放到其他地方,如数据库 Redis、HBase 等。但是若把这些定时检测数据放入 Redis 中,则此时会引发以下问题。

- 这些数据的存放与时间轮刻度如何映射?
- 时间轮存储的检测数据有可能在不断地更新,在时间轮指针每走一刻时,应该如何获取最新的检测数据呢?
- 当时间轮服务宕机或发版重启时,在服务恢复正常后,这些定时检测数据该如何处理,若像 Netty 服务那样,直接把这些数据丢弃,再重写一遍 Redis,则可能会遇到数据严重阻塞,还有丢数据的可能,需要找到其他更好的解决办法。

(1)数据的存放主要考虑获取方便,在获取时,时间轮只需提供当前刻度编号 idx、时间轮唯一标识 wheel、时间轮指针走过了多少刻度 tick 即可,这些数据代表了时间轮的当前状态。若用 HBase 来存储,则可以采用前缀扫描 Scan;若用 Redis 来存储,则可以考虑存放在多个 List 中,这些 List 的 Key 的前缀一致,由 node+idx+tick 组成。先从 Redis 中根据前缀获取这些 Key,再把对应的定时检测数据捞出来。

（2）通过 Key 可以直接获取在时间轮上映射的任务数据，但这些数据早已不再是最新数据了，为了防止误报，需要获取最新数据，此时就需要在这些数据中设置唯一的 id 与最新数据的 id 一致，并把这些原始数据存放在额外的容器中，通过 id 及时覆盖旧的数据。也可以通过 id 链表批量从容器中获取最新数据。

（3）当时间轮所在的服务宕机或重启时，在服务恢复后，只需恢复时间轮的元数据即可，包括其启动时间、指针目前移动了多少刻度、时间轮本身唯一标识、每刻度持续时长、指针走一轮的总刻度值、指针当前所在的刻度编号。根据这些属性重新构建时间轮，无须做任何数据的回放工作。但是要注意，时间轮指针每走一刻度，就需要把时间轮当前的状态进行及时更新，时间轮的状态信息也可以存储到数据库中，如 MySQL、Elasticsearch、HBase、Redis。

经过上述详细分析，有了初步的改造时间轮的方案，下一步需要设计系统的整体架构。

7.3　时间轮高级应用之架构设计

运用 7.2 节时间轮的改造思想，本节会完成一套一天监控 10 亿级定时检测任务系统的架构设计。这套系统的应用范围非常广。例如，外卖平台每天处理上千万份的订单，它是如何确保这些订单在合理的时间内送达的呢？若大量外卖单出现超时未处理问题，那么超过多久后，将无法在预定的时间内送达而收到用户的投诉。若有一套报警系统，当用户下单后，在规定的时间内无任何处理时，能及时给相关人员发送告警信息，操作员收到信息后及时处理，则可以完成订单时效的达成，从而减少用户投诉。这样不仅可以给公司减少赔偿，还对公司的品牌形象有很大的影响。

虽然已经有了明确的改造方案，但由于系统吞吐量平均一分钟要处理上百万条数据，所以在编码开始前，需要进行架构设计、评审。在架构设计前，还需要进行技术选型。技术选型需要考虑以下几点。

- 采用合适的组件作为实时处理数据源。
- 时间轮计算服务需要在分布式组件上运行。

- 数据存储除了 Redis，其他数据库是否也能支持。
- 最终结果数据的输出方案。

（1）由于数据量庞大，而且要时刻监控数据是否被及时处理，因此数据源选择 Kafka 比较合适。当数据处理能力弱时，数据留在 Kafka 中不会丢失，对数据生产者也不会造成影响，而且其扩展性极强，性能也非常不错。

（2）分布式实时流计算框架可以选择 Flink 或 JStorm，它们都具有分布式动态扩展、能接入 Kafka 数据实时流处理、吞吐量高等特点。Flink 不需要对每条消息做 ACK；与 JStorm 相比，它在吞吐量上会更高；有 checkpoint 机制，当 checkpoint 成功时，代表之前的数据都成功的消费了，Kafka 消费组对各分区的消费的偏移量会进行相应的更新。

（3）数据库除了 Redis，还有 HBase。只是 HBase 的批量 GET 偶尔会出现有些数据获取不到的情况，在 region 分裂时可能也会抛异常。本书只提供一个使用 HBase 的思路：当使用 HBase 时，任务数据的 rowkey 前缀与 Redis 的链表 Key 前缀类似，由 wheel+idx+tick 组成。当时间轮每走一格时，可通过 rowkey 前缀循环 Scan 把对应格子中的任务都捞出来并进行处理。

（4）结果输出有两种方案：第一种方案（也是最便捷的方案）是选择 Kafka；第二种方案是通过配置一个 URL http 请求地址来回调给对方。虽然异常数据不会太多，但为了方便，推荐使用 Kafka 方案。感兴趣的话也可以两者都用。

监控 10 亿级定时检测任务系统的架构设计图如图 7-2 所示，图中有多个时间轮，每个时间轮由常量+Flink 的任务 id 组成，由于系统处理的数据量非常大，因此需要两套分布式实时处理程序：第一套加上时间轮，用于任务数据在时间轮上的映射和存储；第二套接收时间轮指针每跳一格发出的消息，根据这些消息从 Redis 数据库中捞取对应格子的任务数据并进行计算，进而输出最终结果。系统可横向无限扩展，整个系统性能瓶颈在 Redis 大批量接收请求上。

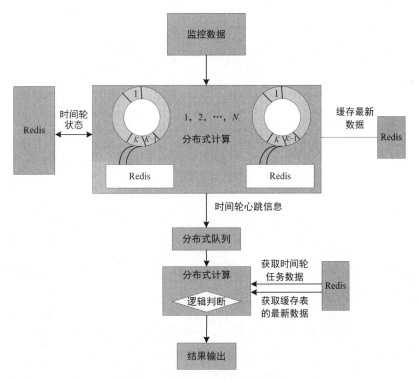

图 7-2　监控 10 亿级定时检测任务系统的架构设计图

7.4　时间轮高级应用之实战 10 亿级任务

本节主要根据 7.3 节的架构设计图来填充数据结构设计，并完成部分代码的编码，数据主要涉及 Kafka 数据源、时间轮状态数据、存储在 Redis 集合中的任务数据，这些数据结构的设计如下。

- 监控数据应该通用化，不只是对于监控定时检测任务。考虑到一般公司都会有调度平台，因此时间轮的高级应用还需要具备周期性定时调度数据的监控。Kafka 的定时监控数据包含以下几个属性：

```
private String dataId;//监控数据id唯一识别
private String sysCode;//系统标识，用于区分是哪个数据源
private long detectionTime;//检测时间
private int isOver;//是否结束检测，1 表示结束，0 表示否
```

```
private long createTm;// 数据创建时间
private int isCycle;//是否需要进行周期监控,0 表示不需要,1 表示需要
private long cycleTime;//监控周期,当 isCycle 为 1 时需要提供
private long opTime;//当前操作时间
private String callBackUrl;//回调 url
```

前面 3 个属性必传,对于其他属性:isOver 与时间轮中的任务取消功能一样;createTm 表示数据生产的时间,用这个时间不仅可以过滤掉一些过期的数据(如数据创建时间大于其检测时间),还可以追踪数据记录;isCycle、cycleTime 与周期调度有关系,只有周期调度任务才需要填充这两个字段;opTime 在计算结果输出时进行设置,结果的输出和 Kafka 数据源可共用同一套数据结构,系统在前期不稳定的情况下可以加上一些其他参数,如时间轮节点名称 wheel、对应的时间轮刻度编号 idx 等,加上这些参数,可以直接从 Redis 中获取对应的数据,进行数据分析,方便定位问题。

- 映射到时间轮刻度上的数据应该具备以下两点:一是可以从最新数据表中获取最新数据;二是可以记录是映射在时间轮的第几轮。因为有些检测延迟时间大于时间轮走一圈的时长。所以这些任务数据存储在 Redis 中,且需要包含以下 3 个属性:

```
private String dataId;//监控数据 id 唯一识别
private String sysCode;//系统标识,用于区分是哪个数据源
private long remainingRounds;//放在时间轮的第几轮
```

- 时间轮的构造参数的定义与业务有一定的关联,若业务一般监控 10 天内超时监测数据,则时间轮的一轮总刻度可以设为 24×10×60。由于量偏大,可以将每刻度持续时长设为 1min。构造参数是时间轮状态的一部分,除了构造参数的基本属性,还需要加上其分配状态、最后执行时间等,一共有以下 7 个属性:

```
private String wheel;//时间轮节点名称
private int flag=1;//1 表示当前时间轮已分配,正在运行;0 表示未分配
private long lastTime;//最后执行时间
private int idx;//当前正在执行的时间轮的刻度
private long startTime;//时间轮开始时间
private long tickDuration;//时间轮每段的间隔时间
private long tick;//时间轮指针走过的刻度
```

系统的核心数据结构已设计完，相信大部分读者可以无须看代码的具体实现逻辑，就有能力完善这套系统的开发了。架构设计是整套系统的核心部分，也是最难的部分。最后需要对 Netty 的时间轮进行以下改造。

- 属性做了变动，删掉了其内存数组、内存泄漏检测、时间轮的内部状态，但同时又新增了时间轮节点属性，并把 Worker 线程中的 tick 移到了时间轮全局属性中，以便其状态恢复。同时加上了缓存 dataCache，用于批量保存数据，减少与 Redis 的 RPC 通信的次数。
- 时间轮构造器需要改造：加上了时间轮的状态属性，以便其状态恢复。
- Worker 线程也需要修改：在 while 循环中，只留下指针等待下一刻度的代码，其他的都可以删掉，同时新增发送消息给 Kafka，消息为 Key 的前缀，由于 1min 内的数据量较大，可考虑把 Key 前缀分成多份，此处可根据业务高峰期进行动态设置，最后需要更新时间轮的状态。
- 在添加任务时，需要把映射到时间轮刻度上的数据的 3 个属性生成 JSON 对象放入缓存 dataCache 中，同时检测 dataCache 数据量是否已经达到阀值，若达到了，则需要将其批量保存在 Redis 中。

对于数据处理端，此处不再赘述了，分布式计算服务在接收到时间轮的定时消息后，从 Redis 中捞取对应的 List 集合，再通过集合中数据的 id 从最新数据表中进行批量获取。需要注意的是，最新数据表的生效日期不能太长，因为数据量非常大，所以需要及时清理过期数据。

系统实现的基本逻辑已完成，具体编码工作由读者自行完成。

7.5 小结

本章主要讲述了时间轮的构建算法及其应用场景。定时检测服务除了外卖平台可以使用，其他地方也可以使用，如优惠券过期提醒、医院预约提醒等，只要是与未来一段时间相关的业务，基本上都可以使用它。

问题分析与性能调优

本章在第 3 章的基础上完成分布式 RPC 服务器的部署、压测、模拟性能瓶颈排查及性能调优。

8.1 Netty 服务在 Linux 服务器上的部署

本书截至目前只有在 Eclipse 或 IntelliJ IDEA 上启动过 Netty 服务,当在本地启动时,一般只用于调试代码;当发布到生产环境中时,需要把服务部署到 Linux 系统上。本节主要对第 3 章的分布式 RPC 代码进行部分修改,因为项目从开发到上线一般会经过开发环境、测试环境、预发环境和生产环境几个环节,且每个环境的配置信息都有所不同,所以本书未选择配置中心,采用的是普通配置文件的方式。每个环境中配置文件的内容都不一样,主要包含

Zookeeper 服务器地址、启动端口等，还需要加上启动 shell 脚本，并用 Java 相关命令把 Netty 服务运行到 Linux 服务器上。

先把服务器名称、启动端口、Zookeeper 服务器地址及服务器权重作为配置参数放到 server.properties 配置文件中。而且在系统测试环境、开发环境、生产环境中各有一份，具体内容如图 8-1 所示。

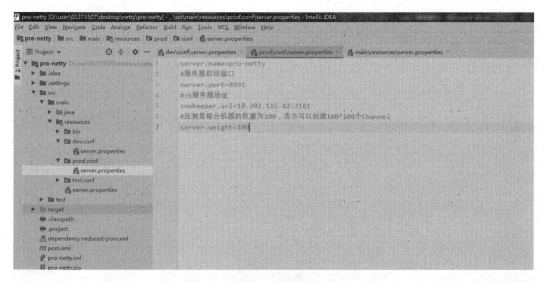

图 8-1　配置文件

这些配置文件需要使用一个工具类 PropertyUtil 来读取，加载配置文件 server.properties 到内存中，并通过 getInt() 与 getString() 方法读取配置属性。编写常量类 Constants，通过工具类获取服务器启动端口和 Zookeeper 服务器地址，具体代码如下：

```
package com.itjoin.pro_netty.util;
import java.util.Properties;
import org.slf4j.Logger;
import org.slf4j.LoggerFactory;
public class PropertyUtil {
    private static Properties serverProperties = new Properties();
    protected static final Logger LOGGER =
                    LoggerFactory.getLogger(PropertyUtil.class);
```

```
    static {
        try {
            serverProperties.load(PropertyUtil.class.
                                    getResourceAsStream("/server.properties"));
            LOGGER.error("加载配置文件成功========");
        } catch (Exception e) {
            e.printStackTrace();
            LOGGER.error("加载配置文件失败========");
        }
    }
    public static String getString(String key) {
        if(serverProperties!=null) {
            return serverProperties.getProperty(key);
        }
        return null;
    }
    public static Integer getInt(String key) {
        if(serverProperties!=null) {
            return Integer.valueOf(serverProperties.getProperty(key));
        }
        return null;
    }
}
public class Constants {
    public static final String SERVER_PATH = "/netty";
    public static int port = 8991;
    public static String zookeeperURL="localhost:2181";
    public static int weight=1;
    static{
        port = PropertyUtil.getInt("server.port");
        zookeeperURL = PropertyUtil.getString("zookeeper.url");
        weight = PropertyUtil.getInt("server.weight");
    }
}
```

代码的改造基本完成了，只剩下权重、Zookeeper 服务器地址及启动端口等常量的替换，读者可自行处理。但还有一个问题：Netty 服务代码如何打包并在 Linux 系统上启动运行。由于 Java 应用服务的启动需要设置 JVM 堆栈内存、lib 包路径及配置文件路径等参数，因此需要编写 shell 启动脚本，步骤如下。

（1）获取 Java 服务所在的目录信息。包含项目依赖的 jar 包、配置文件、日志等目录。

（2）检测 server.properties 中的启动端口是否被占用，若被占用了，则启动失败。

（3）设置 JVM 堆栈内存，采用垃圾回收策略，根据启动参数判断是否要开启 JMX、是否采用调试方式启动服务。

（4）在启动 Java 服务进程后，若当前用户想退出登录，则要使服务进程不结束，此时需要运用 Linux 的后台启动命令 nohup。nohup 命令与 Java 命令配合实现 Netty 服务后台运行在 Linux 系统上。

具体启动脚本 start.sh 代码如下：

```
#!/bin/bash
cd `dirname $0`
BIN_DIR=`pwd`
cd ..
#获取服务包根目录
DEPLOY_DIR=`pwd`
#配置文件目录
CONF_DIR=$DEPLOY_DIR/conf
#日志输出
LOGS_DIR=$DEPLOY_DIR/logs
if [ ! -d $LOGS_DIR ]; then
    mkdir $LOGS_DIR
fi
log_file="$LOGS_DIR/log.start"
if [ -f $log_file ]; then
    rm -rf $log_file
fi
exec 1>>$log_file
#从配置文件中获取服务器名称和启动端口
SERVER_NAME=`sed '/server..name/!d;s/.*=//' conf/server.properties | tr -d '\r'`
```

```
SERVER_PORT=`sed '/server.port/!d;s/.*=//' conf/server.properties | tr -d '\r'`
echo "project name:$SERVER_NAME"
echo "server port:$SERVER_PORT"
if [ -z "$SERVER_NAME" ]; then
    SERVER_NAME=`hostname`
fi
#先判断服务是否已启动
PIDS=`ps aux | grep java | grep "$CONF_DIR" |awk '{print $2}'`
#若服务已启动,则退出
if [ -n "$PIDS" ]; then
    echo "ERROR: The $SERVER_NAME already started!"
    echo "PID: $PIDS"
    exit 1
fi
#判断启动端口是否被占用
if [ -n "$SERVER_PORT" ]; then
    SERVER_PORT_COUNT=`netstat -tln | grep "$SERVER_PORT" | wc -l`
    if [ $SERVER_PORT_COUNT -gt 0 ]; then
        echo "ERROR: The $SERVER_NAME port $SERVER_PORT already used!"
        exit 1
    fi
fi
LOGS_DIR=$DEPLOY_DIR/logs
if [ ! -d $LOGS_DIR ]; then
    mkdir $LOGS_DIR
fi
#控制台日志输出
STDOUT_FILE=$LOGS_DIR/stdout.log
#jar 包目录,包含第三方依赖 jar
LIB_DIR=$DEPLOY_DIR/lib
#获取所有的 jar 包路径,用":"连接
LIB_JARS=`ls $LIB_DIR|grep .jar|awk '{print "'$LIB_DIR'/"$0}'|tr "\n" ":"`
JAVA_OPTS=" -Djava.awt.headless=true -Djava.net.preferIPv4Stack=true "
JAVA_DEBUG_OPTS=""
if [ "$1" = "debug" ]; then
    JAVA_DEBUG_OPTS=" -Xdebug -Xnoagent -Djava.compiler=NONE -Xrunjdwp:transport=dt_socket,address=8000,server=y,suspend=n "
```

```
fi
JAVA_JMX_OPTS=""
#是否开启jmx
if [ "$1" = "jmx" ]; then
    JAVA_JMX_OPTS="  -Dcom.sun.management.jmxremote.port=1099  -Dcom.sun.management.jmxremote.ssl=false -Dcom.sun.management.jmxremote.authenticate=false "
fi
#根据jdk版本设置堆栈内存
JAVA_MEM_OPTS=""
BITS=`java -version 2>&1 | grep -i 64-bit`
if [ -n "$BITS" ]; then
    JAVA_MEM_OPTS="  -server  -Xmx8g  -Xms8g  -Xmn512m  -XX:PermSize=128m  -Xss256k -XX:+DisableExplicitGC -XX:+UseConcMarkSweepGC -XX:+CMSParallelRemarkEnabled -XX:+UseCMSCompactAtFullCollection -XX:LargePageSizeInBytes=128m -XX:+UseFastAccessorMethods -XX:+UseCMSInitiatingOccupancyOnly -XX:CMSInitiatingOccupancyFraction=70 "
else
    JAVA_MEM_OPTS=" -server -Xms1g -Xmx1g -XX:PermSize=128m -XX:SurvivorRatio=2 -XX:+UseParallelGC -Xloggc:$LOGS_DIR/gc.log"
fi
echo -e "Starting the $SERVER_NAME ...\c"
params=$1
#hohup命令后台启动
nohup java $JAVA_OPTS $JAVA_MEM_OPTS $JAVA_DEBUG_OPTS $JAVA_JMX_OPTS -classpath $CONF_DIR:$LIB_JARS com.itjoin.pro_netty.spring.ApplicationMain $params >$STDOUT_FILE 2>&1 &
echo "wait the server starting..."
#判断是否成功启动,并输出启动日志
COUNT=0
while [ $COUNT -lt 1 ]; do
    echo -e ".\c"
    sleep 1
    COUNT=`ps aux | grep java | grep "$DEPLOY_DIR" | awk '{print $2}' | wc -l`
    if [ $COUNT -gt 0 ]; then
        break
    fi
done
echo "OK!"
PIDS=`ps aux | grep java | grep "$DEPLOY_DIR" | awk '{print $2}'`
```

```
echo "PID: $PIDS"
echo "STDOUT: $STDOUT_FILE"
```

有了启动脚本,再把所有依赖的 jar 包和配置文件打成 zip 包,并上传到 Linux 服务上。然后解压文件,运行 start.sh 脚本即可。

Maven 提供了非常好的 zip 打包插件 maven-assembly-plugin。此插件需要提供一个 XML 配置文件,一般取名为 assembly.xml。assembly.xml 文件的内容会根据 Maven 的 profile 文件来解决不同环境中配置文件的部署问题。接下来在 pom.xml 文件中添加此插件代码和各环境下 profile 的属性。具体代码如下:

```xml
<profiles>
    <!--设置好各个环境的 env 属性值,打包环境由打包命令-P 参数指定-->
    <profile>
        <id>dev</id>
        <activation>
        </activation>
        <properties>
            <env>dev</env>
        </properties>
    </profile>
    <profile>
        <id>prod</id>
        <activation>
        </activation>
        <properties>
            <env>prod</env>
        </properties>
    </profile>
    <profile>
        <id>test</id>
        <activation>
        </activation>
        <properties>
            <env>test</env>
        </properties>
```

```xml
        </profile>
</profiles>
<build>
    <plugins>
        <!--只在第一次打 zip 包时打开这个插件-->
        <plugin>
            <groupId>org.apache.maven.plugins</groupId>
            <artifactId>maven-assembly-plugin</artifactId>
            <version>2.2.1</version>
            <configuration>
                <descriptors>
<descriptor>src/main/resources/bin/assembly.xml</descriptor>
                </descriptors>
            </configuration>
            <executions>
                <execution>
                    <id>make-assembly</id>
                    <phase>package</phase>
                    <goals>
                        <goal>single</goal>
                    </goals>
                </execution>
            </executions>
        </plugin>
</plugins>
</build>
```

assembly.xml 代码如下:

```xml
<assembly>
    <id>${env}</id>
    <formats>
        <format>zip</format>
    </formats>
    <includeBaseDirectory>true</includeBaseDirectory>
    <fileSets>
        <fileSet>
            <directory>src/main/resources/bin</directory>
```

```xml
            <outputDirectory>bin</outputDirectory>
            <directoryMode>0755</directoryMode>
            <fileMode>0755</fileMode>
        </fileSet>
        <fileSet>
            <!--${env}为pom.xml profile中根据-P参数获取打包环境的属性env的值-->
            <directory>src/main/resources/${env}/conf</directory>
            <outputDirectory>conf</outputDirectory>
            <directoryMode>0744</directoryMode>
            <fileMode>0644</fileMode>
        </fileSet>
        <fileSet>
            <directory>lib</directory>
            <outputDirectory>lib</outputDirectory>
            <directoryMode>0744</directoryMode>
            <fileMode>0644</fileMode>
        </fileSet>
    </fileSets>
    <dependencySets>
        <dependencySet>
            <outputDirectory>lib</outputDirectory>
        </dependencySet>
    </dependencySets>
</assembly>
```

最后运行 Maven 命令":mvn clean package-DskipTests-Psit"，-P 参数表示指定具体 profile 环境。运行成功后，在 target 目录下会生成一个名为 pro-netty-0.0.1-SNAPSHOT-test.zip 的文件。用终端模拟软件 Xshell 连接 Linux 系统，把 zip 包上传并解压，再运行 start.sh 脚本启动。若出现/bin/bash^M:异常，则可能是编码的问题，需要进行编码转换。此时用 vi 命令输入":set ff=unix"，更改其格式，然后保存退出，就能正常运行了，至此服务器部署完成。需要注意的是，在启动 Netty 服务前，需要保证 Zookeeper 服务器处于运行状态。

8.2　Netty 服务模拟秒杀压测

系统在正式上线前，一般会根据系统的用户数、预估接口的调用量等信息对预发布系统进行压测。Netty 分布式服务性能究竟如何，大概需要多少资源才能撑住网络购物高峰期？由于没有任何性能指标数据，无法评估，所以只能通过压测获取服务器的 TPS、并发用户数、平均响应耗时等数据，才能对服务的处理能力做出精准的预估。本节模拟秒杀活动逻辑，对 Netty 分布式服务进行压测。在压测过程中能发现系统不稳定的地方及其性能瓶颈，然后可以采用各种调优方案尝试对系统进行优化。

Jmeter 是比较常用的压测工具，它有界面化配置，支持模拟发送 HTTP、TCP、UDP、FTP 等请求包，一般都会选择它作为微服务的压测工具。通过设置好请求的配置信息运行 Jmeter 压测脚本，即可进行压测。那么，该如何运用 Jmeter 对 Netty 分布式服务进行压测呢？

采用普通 Jmeter TCP 压测脚本，只能压测其中一台服务器。此时，读者可能会想到运用 Nginx 或 LVS 等反向代理、负载均衡软件进行转发，但这种方式需要加入 Nginx 服务器资源。另一种方式是采用 Jmeter 对 jar 包的嵌入调用来进行模拟测试，这也是本书选择的压测方式。从 Jmeter 官网下载 apache-jmeter-5.1.1.zip，解压后，进入 bin 目录，双击 jmeter.bat 脚本即可启动。在启动 Jmeter 之前，需要把调用的 jar 包准备好。

在打包之前，需要修改 pom.xml 文件。在文件中需要把 Jmeter 的 lib/ext 目录下的 ApacheJMeter_java.jar 文件依赖进去。这个 jar 包在 Maven 仓库中无法下载，需要从 Jmeter 对应的目录复制到本地仓库中。同时，需要添加一个打包插件，此插件和 8.1 节中的插件不同，它会把依赖 jar 包的所有类全部打到同一个包里。

```xml
<!--此依赖需要把 Jmeter 安装目录下的 lib/ext 目录中的 ApacheJMeter_java.jar 复制到本地仓
库中-->
<dependency>
    <groupId>org.apache.jmeter</groupId>
    <artifactId>ApacheJMeter_java</artifactId>
    <version>5.1.1</version>
    <scope>provided</scope>
```

```xml
            <exclusions>
                <exclusion>
                    <groupId>org.apache.logging.log4j</groupId>
                    <artifactId>log4j-api</artifactId>
                </exclusion>
                <exclusion>
                    <groupId>org.apache.logging.log4j</groupId>
                    <artifactId>log4j-slf4j-impl</artifactId>
                </exclusion>
                <exclusion>
                    <groupId>org.apache.logging.log4j</groupId>
                    <artifactId>log4j-core</artifactId>
                </exclusion>
                <exclusion>
                    <groupId>org.apache.logging.log4j</groupId>
                    <artifactId>log4j-1.2-api</artifactId>
                </exclusion>
            </exclusions>
        </dependency>
    </dependencies>
    <build>
        <plugins>
            <!--打完Jmeter包后需要注释这个插件-->
            <plugin>
                <groupId>org.apache.maven.plugins</groupId>
                <artifactId>maven-shade-plugin</artifactId>
                <version>2.4.1</version>
                <executions>
                    <execution>
                        <phase>package</phase>
                        <goals>
                            <goal>shade</goal>
                        </goals>
                        <configuration>
                            <transformers>
                                <transformer implementation="org.apache.maven.plugins.shade.resource.ManifestResourceTransformer">
```

第 8 章 问题分析与性能调优

```xml
<!--下面这个类是Jmeter调用类的入口,打包时需要把所有的依赖jar包全部打进去-->
<mainClass>com.itjoin.pro_netty.jmeter.RPCJmeterClient</mainClass>
                    </transformer>
                </transformers>
            </configuration>
        </execution>
    </executions>
  </plugin>
 </plugins>
</build>
```

新建一个测试入口类,取名为 RPCJmeterClient,它需要继承 Jmeter 包中的抽象类——org.apache.jmeter.protocol.java.sampler.AbstractJavaSamplerClient,并重写此抽象类的 4 个方法:getDefaultParameters()、setupTest()、runTest()和 teardownTest()。这 4 个方法的主要功能有请求参数初始化、请求的并发调用入口及压测完的收尾工作,具体说明如下:

```java
//设置可用参数,默认为null
public Arguments getDefaultParameters();
//每个线程测试前执行一次,做一些初始化工作
public void setupTest(JavaSamplerContext arg0);
//开始测试,从arg0参数可以获得请求参数值
public SampleResult runTest(JavaSamplerContext arg0);
//测试结束时调用
public void teardownTest(JavaSamplerContext arg0);
```

RPCJmeterClient 的主要实现逻辑:初始化 Spring 容器,获取用于压测的 LoginController 对象,并调用其压测方法。在秒杀场景中,一般会用到 Redis 缓存数据库,但由于服务器资源有限,只用了本地缓存进行模拟,感兴趣的读者可自行完成秒杀逻辑的修改。详细的实现代码如下:

```java
package com.itjoin.pro_netty.jmeter;
import com.alibaba.fastjson.JSONObject;
import com.itjoin.pro_netty.controller.LoginController;
import org.apache.jmeter.config.Arguments;
import org.apache.jmeter.protocol.java.sampler.AbstractJavaSamplerClient;
import org.apache.jmeter.protocol.java.sampler.JavaSamplerContext;
```

```java
import org.apache.jmeter.samplers.SampleResult;
import org.springframework.context.annotation.AnnotationConfigApplicationContext;
public class RPCJmeterClient extends AbstractJavaSamplerClient{
    //原子性初始化
    public static AtomicBoolean hasInit=new AtomicBoolean(false);
    /**
     * 参数信息,此处可以通过Arguments的addArgument()方法加上请求参数
     * @return
     */
    @Override
    public Arguments getDefaultParameters() {
        return null;
    }
        //需要静态化,与springmvc的Controller默认单例相似
    static LoginController loginController;
    /**
     * 开始测试
     * @param context
     * @return
     */
    @Override
    public SampleResult runTest(JavaSamplerContext context) {
        SampleResult result = new SampleResult();
        try {
            //开始计时,此处一定要记得设置,否则看不到TPS
            result.sampleStart();
            //请求分布式RPC服务器
            Object response = loginController.testSecondSell("1");
                result.setResponseData(JSONObject.toJSONString(response),"utf-8");
            result.setResponseOK();
            //计时结束
            result.sampleEnd();
            //响应正常
            result.setSuccessful(true);
        } catch (Exception e) {
            e.printStackTrace();
            //响应异常
```

```java
        result.setSuccessful(false);
    }
    return result;
}
/**
 * 每个线程测试前执行一次，做一些初始化工作
 * @param context
 */
@Override
public void setupTest(JavaSamplerContext context) {
    super.setupTest(context);
    //初始化Spring容器（与TestProxyRpc的main()方法的内容一致）
    //加锁是为了防止多次初始化
    synchronized (hasInit){
        if(!hasInit.getAndSet(true)){
            System.out.println("启动容器初始化===");
            AnnotationConfigApplicationContext springContext = new
                AnnotationConfigApplicationContext(
                new String[]{"com.itjoin.pro_netty.controller",
                    "com.itjoin.pro_netty.proxy"});
            loginController= springContext.
                getBean(LoginController.class);
            //socket连接初始化,防止多线程一起调用
            try {
                ServerChangeWatcher.initChannelFuture();
            } catch (Exception e) {
                e.printStackTrace();
            }
        }
    }
}
/**
 * 压测结束时被调用
 * @param context
 */
public void teardownTest(JavaSamplerContext context) {
    System.out.println("======over====");
}
```

```
}
//loginController 类、UserService 接口和 UserServiceImpl 类需要加上 testSecondSell()方法
//loginController 类加上 testSecondSell()方法
    public Object testSecondSell(String productId){
    return userService.testSecondSell(productId);
}
    //新增 UserService 接口
    public Object testSecondSell(String productId);
    //新增 UserServiceImpl 类
    public static Map<String,Integer> products = new ConcurrentHashMap<>();
static{
    products.put("1",Integer.MAX_VALUE);
}
/**
 * 秒杀活动测试
 * @param productId
 * @return
 */
@Override
public Object testSecondSell(String productId) {
    //此处需改成 Redis 分布式锁,由于无 Redis 资源,所以只采用了 JVM 锁
    synchronized (products){
        //此处库存也放在本地缓存中,正常秒杀应该放入 Redis 中
        Integer productVal = products.get(productId);
        if(productVal>0){
            System.out.println("ok");
            products.put(productId,productVal-1);
            return "购买成功";
        }
    }
    return "购买失败";
}
```

最后,需要打包并设置好 Jmeter 配置脚本。执行 Maven 打包命令 mvn clean package –DskipTests,把完整包(包括所有依赖的 jar 包)放在 apache-jmeter-5.1.1\lib\ext 目录下,启动 jmeter.bat 脚本,在"TestPlan"目录下新建 Thread Group,如图 8-2 所示。

在图 8-2 中，30000 表示压测用户线程数，10（单位为 s）表示线程完成时间。右击"ThreadGroup"目录，在弹出的快捷菜单中依次选择"Add"|"Sampler"|"Java Request"选项，进入 JavaRequest 界面，选择 Classname 为 com.itjoin.pro_netty.jmeter.RPCJmeterClient。最后，右击"Java Request"目录，在弹出的快捷菜单中，依次选择"Add"|"Listener"|"Aggregate Report"选项，新增聚合报告，可以关注其响应平均耗时、错误率及吞吐量等，如图 8-3 所示。

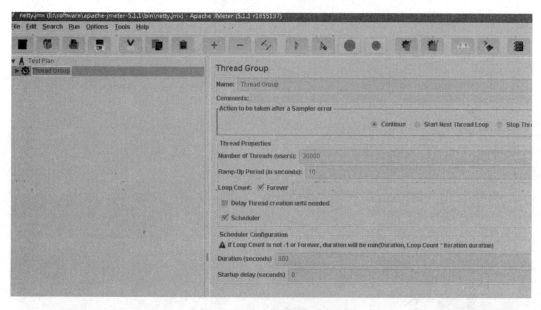

图 8-2　Jmeter 线程组配置

图 8-3　Jmeter 聚合报告界面

将压测脚本保存并命名为 Netty.jmx，上传到 Linux 物理机上的 Jmeter 的 bin 目录下，运行 Jmeter 压测命令：sh ./bin/jmeter -n -t netty.jmx -l netty_result.jtl -e -o /tmp/netty_yc，Jmeter 在运行时可能会

抛出内存溢出异常，这时需要调整 JVM 内存，把 Jmeter 文件中的${HEAP:="-Xms1g -Xmx1g -XX:MaxMetaspaceSize=256m"}改成${HEAP:="-Xms8g -Xmx8g -XX: MaxMetaspaceSize = 512m"}；把 1GB 的内存调整到 8GB，再次运行。

在压测时，采用 3 台 20 核超线程、内存为 256GB 的服务器。一台用来部署 Jmeter 压测机，另外两台用来部署 2 个 Netty 服务节点。机器上还装了多个很耗性能的 Elasticsearch 节点，对压测有一定的影响。为了防止服务器宕机，并没有达到极限性能。在 Netty 服务上，Worker 线程数默认为 CPU 的 2 倍，其压测结果如图 8-4 所示，平均 TPS 为 114147 次/s。

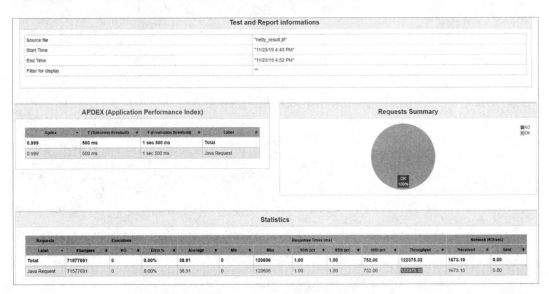

图 8-4　2 台服务器 30000 线程压测报告

从压测结果可以看出，Jmeter 的压测线程数并不是在 10s 时就全部启动完成的，这主要是因为只用了一台 Jmeter 服务器，启动 30 000 条线程后，通过图 8-5 可以看出，Jmeter 服务器 CPU 使用了 2278%，整个 CPU 使用了 34.8%。图 8-6 显示的是 Netty 服务器 CPU 的使用情况，当前 Netty 服务器 CPU 使用了 461.5%，整个 CPU 使用了 5.9%。Netty 服务器 CPU 的使用率没提上去，说明增加机器的效果不会太好。30 000 条线程对于单台 Jmeter 的压力也并不算太大，其主要问题是创建 30 000 个 TCP 长连接。最有效的办法是增加 Jmeter 服务器，采用 Jmeter

集群压测，同时稍微增加 Netty 客户端与服务器的连接数。但注意不要过多，以防内存溢出。

本节的压测只是给读者提供了一个压测案例，并没有进行过多的性能调优。在真实压测场景中包含复杂的业务逻辑，肯定会遇到各种问题，性能可能一直无法达标，此时可借用工具查找其性能瓶颈，进而想办法优化。若在代码中未发现性能瓶颈，则可以采用调整线程参数、扩充服务器等方法。

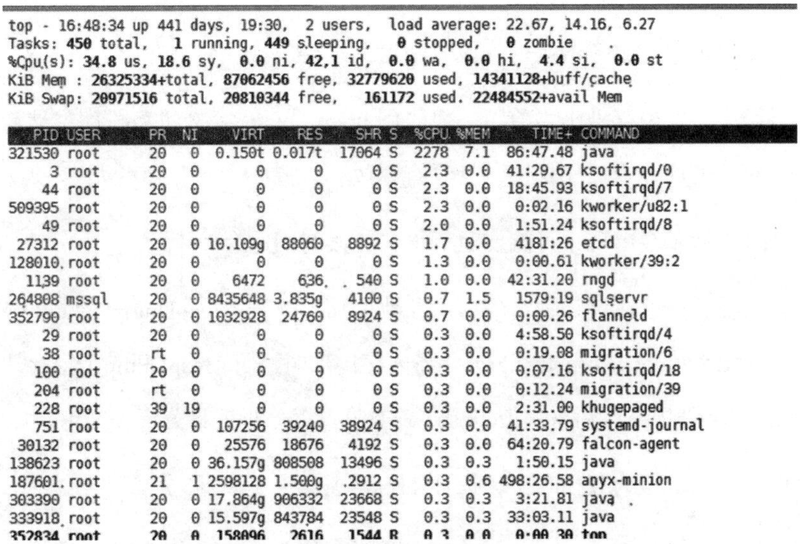

图 8-5　压测时 Jmeter 服务器 CPU 的使用情况

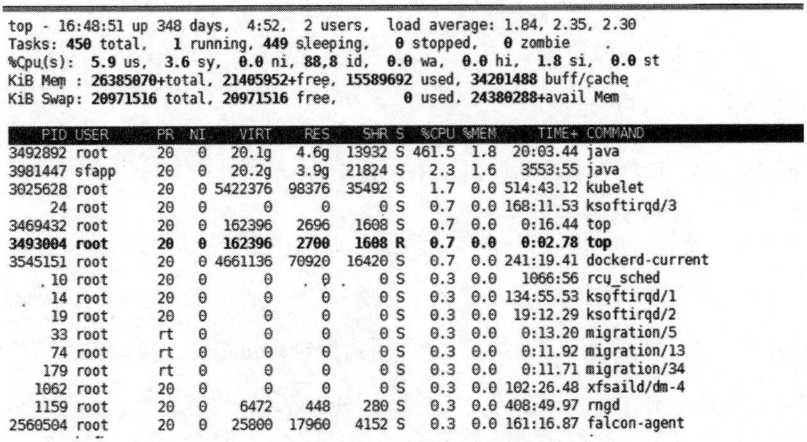

图 8-6　压测时 Netty 服务器 CPU 的使用情况

8.3 常见生产问题分析

虽然压测 Netty 服务器的计算逻辑非常简单，但在实际应用中，在 Service 类里会有很多复杂逻辑，包括数据库操作、远程服务的调用等。若在压测过程中，服务器性能一直无法达标，则会经常出现内存溢出的情况。这时候该如何处理呢？下面列出了几种常见问题及其处理方式。

（1）内存溢出现象在服务器运行一段时间后一定会出现。

针对这种情况，在内存溢出之前，一般先用 jmap –histo 命令进行排序，然后观察一段时间，看看哪些对象的内存增长比较快，且没有释放。

（2）内存溢出偶尔会出现，可能要运行很长一段时间才会发生。

针对这种情况，一般在启动脚本的启动参数上加上-XX:+HeapDumpOnOutOfMemoryError -XX:HeapDumpPath=/tmp/ pro-netty.hprof。当内存溢出时进行 HeapDump（内存快照）并保存到 pro-netty.hprof 文件中，然后采用 Eclipse MAT 等内存分析工具分析该文件。

（3）无内存溢出，但响应耗时比较长且 TPS 非常低。

针对这种情况，采用 top –H –p pid 命令找到第一条线程 id，把线程 id 转换成十六进制小写的形式，运行 "jstack –l pid | grep '线程 id16 进制小写'"。找到占用 CPU 比较高的线程的运行栈信息，但这种方式不够灵活，无法检索具体耗时、超过了多少代码。推荐使用阿里巴巴的工具 Arthas 的 trace 命令进行定位。

（4）无内存溢出，但发现有些功能出现卡死现象。

例如，Flink 任务的 checkpoint 一直无法成功，此时该如何定位呢？可以通过 Arthas 工具的 thread 命令来查看当前线程的情况。如图 8-7 所示，Flink 由于处理端的消费能力太弱，导致数据生产端 Source 线程获取不到内存，却一直占用锁，每等待一段时间就会检测是否有内存。若一直获取不到，则一直循环等待。这会导致 checkpoint 线程阻塞，无法获取锁，也无法向其下游广播 barrier 信号。在图 8-7 中，线程 77 和线程 81 获取的锁对象是同一个，由于线程

77 一直在循环等待，导致线程 81 阻塞，形成了第 2 章中讲的背压。

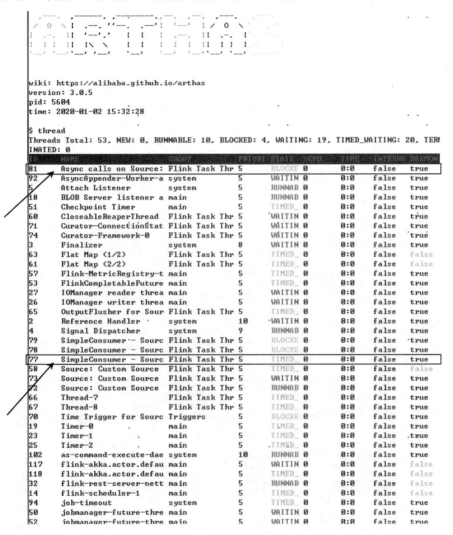

图 8-7　Flink 服务线程运行情况

在图 8-8 中，77 线程获取到锁 java.lang.Object@39820a03，但由于没有足够的内存，无法给下游发送数据，只能不断地等待内存释放。而在图 8-9 中，81 线程与 78 线程都处于 blocked 状态，等待线程 77 释放 java.lang.Object@39820a03 对象。此处主要用到了 Arthas 的 thread 命令（查看线程栈的具体情况）。至于 Arthas 的其他功能，读者可以参考其 GitHub。

```
                at org.apache.flink.streaming.connectors.kafka.internals.AbstractFetcher.emitRecord(A
        bstractFetcher.java:361)
                - blocked on java.lang.Object@39820a03
                at org.apache.flink.streaming.connectors.kafka.internals.SimpleConsumerThread.run(Sim
        pleConsumerThread.java:381)
        Affect(row-cnt:0) cost in 33 ms.
        $ thread 77
        "SimpleConsumer - Source: Custom Source - broker-3 (10.206.216.7:9092)" Id=77 TIMED_WAITI
        NG on java.util.ArrayDeque@94f313d
                at java.lang.Object.wait(Native Method)
                - waiting on java.util.ArrayDeque@94f313d
                at org.apache.flink.runtime.io.network.buffer.LocalBufferPool.requestMemorySegment(Lo
        calBufferPool.java:240)
                at org.apache.flink.runtime.io.network.buffer.LocalBufferPool.requestBufferBuilderBlo
        cking(LocalBufferPool.java:197)
                at org.apache.flink.runtime.io.network.api.writer.RecordWriter.requestNewBufferBuilde
        r(RecordWriter.java:209)
                at org.apache.flink.runtime.io.network.api.writer.RecordWriter.sendToTarget(RecordWri
        ter.java:142)
                at org.apache.flink.runtime.io.network.api.writer.RecordWriter.emit(RecordWriter.java
        :105)
                at org.apache.flink.streaming.runtime.io.StreamRecordWriter.emit(StreamRecordWriter.j
        ava:81)
                at org.apache.flink.streaming.runtime.io.RecordWriterOutput.pushToRecordWriter(Record
        WriterOutput.java:107)
                at org.apache.flink.streaming.runtime.io.RecordWriterOutput.collect(RecordWriterOutpu
        t.java:89)
                at org.apache.flink.streaming.runtime.io.RecordWriterOutput.collect(RecordWriterOutpu
        t.java:45)
                at org.apache.flink.streaming.api.operators.AbstractStreamOperator$CountingOutput.col
        lect(AbstractStreamOperator.java:679)
                at org.apache.flink.streaming.api.operators.AbstractStreamOperator$CountingOutput.col
        lect(AbstractStreamOperator.java:657)
                at org.apache.flink.streaming.api.operators.StreamSourceContexts$NonTimestampContext.
        collect(StreamSourceContexts.java:104)
                - locked java.lang.Object@39820a03
                at org.apache.flink.streaming.connectors.kafka.internals.AbstractFetcher.emitRecord(A
        bstractFetcher.java:362)
                - locked java.lang.Object@39820a03
                at org.apache.flink.streaming.connectors.kafka.internals.SimpleConsumerThread.run(Sim
        pleConsumerThread.java:381)
        Affect(row-cnt:0) cost in 35 ms.
```

图 8-8　Flink 77 线程等待获取内存

```
        $ thread 81
        "Async calls on Source: Custom Source (1/1)" Id=81 BLOCKED on java.lang.Object@39820a03 o
        wned by "SimpleConsumer - Source: Custom Source - broker-3 (10.206.216.7:9092)" Id=77
                at org.apache.flink.streaming.runtime.tasks.StreamTask.performCheckpoint(StreamTask.j
        ava:621)
                - blocked on java.lang.Object@39820a03
                at org.apache.flink.streaming.runtime.tasks.StreamTask.triggerCheckpoint(StreamTask.j
        ava:564)
                at org.apache.flink.streaming.runtime.tasks.SourceStreamTask.triggerCheckpoint(Source
        StreamTask.java:116)
                at org.apache.flink.runtime.taskmanager.Task$2.run(Task.java:1210)
                at java.util.concurrent.Executors$RunnableAdapter.call(Executors.java:511)
                at java.util.concurrent.FutureTask.run$$capture(FutureTask.java:266)
                at java.util.concurrent.FutureTask.run(FutureTask.java)
                at java.util.concurrent.ThreadPoolExecutor.runWorker(ThreadPoolExecutor.java:1149)
                at java.util.concurrent.ThreadPoolExecutor$Worker.run(ThreadPoolExecutor.java:624)
                at java.lang.Thread.run(Thread.java:748)

                Number of locked synchronizers = 1
                - java.util.concurrent.ThreadPoolExecutor$Worker@2f7a0400

        Affect(row-cnt:0) cost in 33 ms.
        $ thread 78
        "SimpleConsumer - Source: Custom Source - broker-2 (10.206.216.12:9092)" Id=78 BLOCKED on
         java.lang.Object@39820a03 owned by "SimpleConsumer - Source: Custom Source - broker-3 (1
        0.206.216.7:9092)" Id=77
                at org.apache.flink.streaming.connectors.kafka.internals.AbstractFetcher.emitRecord(A
        bstractFetcher.java:361)
                - blocked on java.lang.Object@39820a03
                at org.apache.flink.streaming.connectors.kafka.internals.SimpleConsumerThread.run(Sim
        pleConsumerThread.java:381)
        Affect(row-cnt:0) cost in 33 ms.
```

图 8-9　Flink 阻塞 81 线程与 78 线程的运行情况

8.4 性能调优

通过 Jmeter 的压测结果可以看出，在无复杂业务逻辑的情况下，Netty 无须调优，其整个 RPC 的性能也表现得很不错。但在真实场景下，一般都会运用各种调优方法进行不断的尝试。例如，学习 Netty 的内存池管理，在对其底层实现有了全面的了解后，可以大胆地运用其内存池来减少 I/O 的读/写堆外内存的分配次数；心跳检测可尝试使用第 7 章的时间轮来替换 Netty 自带的检测机制，减轻 I/O 线程负担。性能调优主要有以下三大方向。

（1）Linux 系统参数调整。

（2）TCP 参数调整。

（3）Netty 服务器应用层优化。

1. Linux 系统参数调整

在压测时，可能会遇到文件句柄数过多异常，此时可以通过修改 Linux 单个进程打开的文件句柄最大值参数来解决。先采用 cat /proc/sys/fs/file-max 命令查看全局文件句柄限制，若偏小，则可通过 root 账号执行 vim /etc/sysctl.conf；如果要支持百万连接，则需要在末尾加上 fs.file-max=1000000。修改完成后，执行 sysctl–p，使其修改生效。同时需要运行 ulimit–n 查看局部文件句柄限制，若偏小，则执行 vim /etc/security/limits.conf，在末尾加上 "*hard nofile 1000000" 和 "*soft nofile 1000000"，保存后重启服务器。

2. TCP 参数调整

（1）TCP 层面的接收和发送缓冲区的大小设置分别对应 Netty 中 ChannelOption 类的 SO_SNDBUF 和 SO_RCVBUF，缓冲区大小一般设为网络吞吐量达到带宽上限时的值，即缓冲区大小=网络带宽×网络时延。以千兆网卡为例进行计算，假设网络时延为 1ms，则缓冲区大小=1000MB/s×1ms = 128KB。

（2）当 TCP 的 keepalive 心跳机制探测到 socket 不可用时，会触发 channelInactive()方法，keepalive 不允许为探测套接字终点指定一个值，而且默认空闲检测时间为 2 个小时，需要修改

Linux 配置，依赖操作系统很不灵活，对 Netty 应用程序没什么用，可以关闭，Netty 用 IdleStateHandler 做心跳检测，但注意检测时间不要过长，一般在几十秒内，Netty 还可采用时间轮做心跳检测以管理链路。

（3）是否复用处于 TIME_WAIT 状态连接的端口，Netty 配置了 ChannelOption.SO_REUSEADDR，当将其设为 true 时，表示复用。当 Netty 服务器调用其他接口或数据库时，若出现大量连接处于 TIME_WAIT 状态的情况，则设置此参数比较有用。

（4）在 TCP 的 3 次握手中，Linux 内核维护了 SYN 队列和 Accept 队列。SYN 队列用于保存半连接状态的请求，当客户端给服务器发送 SYN 后，TCP 连接的状态变成 SYN_SEND；当服务器收到请求后，将其状态变成 SYN_RCVD，并把该请求放入 SYN 队列中。此队列的大小可通过 Netty 的参数 ChannelOption.SO_BACKLOG 来设置，其大小一定要小于 ulimit -n 返回的值。

当 3 次握手完成后，请求会从 SYN 队列中被移到 Accept 队列中。有一点要注意，TCP SYN FLOOD 恶意 DOS 攻击方式会创建大量的半连接状态的请求，请求的源地址是伪造的，服务器将会消耗非常多的资源来处理这种半连接，而且会使其他正常请求无法保存到此队列中。Accept 队列中存放的是已经建立好的连接，等待 Netty 的 Boss 线程把它们取走。

TCP 参数优化代码如下：

```
serverBootstrap.childOption(ChannelOption.SO_SNDBUF,128*1024);
serverBootstrap.childOption(ChannelOption.SO_RCVBUF,128*1024);
serverBootstrap.childOption(ChannelOption.SO_KEEPALIVE,false);
serverBootstrap.childOption(ChannelOption.SO_REUSEADDR,true);
// cat /proc/sys/net/ipv4/tcp_max_syn_backlog 可查看其默认值
serverBootstrap.option(ChannelOption.SO_BACKLOG, 2048);
```

3. Netty 服务器应用层优化

（1）线程调优，在压测 Netty 分布式 RPC 服务时，除 Boss 线程外，只开了 I/O Worker 线程。在长连接情况下，当 Worker 线程调到用户连接数时，继续增大 Worker 线程，性能也无法再提升。主要是 I/O 线程受到了睡眠阻塞，此时需要开启业务线程池处理业务，不要阻塞 I/O 线程，性能会得到一定的提升。在使用线程池时，不要用 JDK 的 Executors 直接 newFixedThreadPool 或 newCachedThreadPool，这两个线程池都可能会导致内存溢出。第一个线程池看似固定了线程数，但其队列是 LinkedBlockingQueue，是无界的。当任务数突然暴增时，排队的任务非常多，有可能会出现内存溢出。第二个队列的线程数上限为 Integer.MAX_VALUE，也很有可能出现内存溢出。

可以直接选择新建一个 ThreadPoolExecutor，根据压测结果，将其最大线程数调到最优。同时新建一个有长度限制的队列 ArrayBlockingQueue<Runnable>(200)。当出现任务无法加入队列的情况时，要触发 Netty 的背压机制，防止服务器被压爆。

（2）JVM 参数。

- -Xms 和 -Xmx：初始堆内存和最大堆内存需要根据内存模型进行计算并得出相对合理的值，并不是内存分配得越大越好。JVM 在内存小于 32GB 的时候，会采用一个内存对象指针压缩技术。此指针不再表示对象在内存中的精确位置，而是表示偏移量。这意味着 32bit 的指针可以引用 40 亿个对象，而不是 40 亿个字节。也就是说，堆内存长到 32GB 的物理内存，也可以用 32bit 的指针来表示。JVM 内存设置一旦超过 30～32GB 的边界，指针就会切回普通对象的指针，每个对象的指针都变长了，就会使用更多的 CPU 内存，实际上失去了更多的内存。事实上，当内存到达 40～50GB 时，真正有效的内存才相当于使用内存对象指针压缩技术时的 32GB 内存。尽量不要超过 32GB，因为它浪费了内存、降低了 CPU 的性能，还要让垃圾回收器应对大内存。当必须要使用大内存时，在垃圾回收器的选择上需要进行更多的考虑。

- -Xss256K：设置每个线程的堆栈大小，JDK1.5 之后的版本都默认为 1MB，根据应用线程所需的内存大小进行调整，在内存相同的情况下，减小这个值能生成更多的线程，但太小容易出现栈溢出异常。

- -XX:NewRatio=4：年轻代与老年代的比值为 4，年轻代占整个堆栈的 1/5。

- -XX:SurvivorRatio=4：将年轻代中的 Eden 区与 Survivor 区的大小比值设置为 4，即一个 Survivor 区占整个年轻代的 1/6。

- -XX:MetaspaceSize=512m：元数据空间大小，主要存储类和方法信息，JDK1.8 之前的版本为-XX:MaxPermSize，称为永久代，此空间基本上不会有太大的改变，除非系统在运行中动态生成大量的类。

JVM 参数调整主要是为了降低 Full GC 的频率，让系统运行更稳定。当堆内存非常大时，垃圾回收器 G1 是比较好的选择。尤其是 CMS 垃圾回收器产生大量内存空间碎片，经常导致年轻代晋升至老年代没有足够的空间，当 Full GC 的频率比较高时，由于 G1 不会产生内存空间碎片，所以换成 G1 后，Full GC 频率可能会急速下降。

（3）对于非实时性特别高的系统，Netty 的写操作尽量减少直接往网络中写的次数，可减少系统调用的开销，提高带宽利用率，如 Flink、JStorm 等流式计算框架。

（4）业务代码优化，如加缓存，一般业务系统接口会加上 Redis 缓存，以减少磁盘的 I/O 操作。

8.5 小结

本章主要讲解了服务器代码的部署、Jmeter 压测、线上问题定位及解决的实战经验、性能调优的方案。服务器代码的部署也是应用开发过程中很重要的部分，对于未接触过 Netty 的读者有一定的帮助。一般开发人员平时很少接触 Jmeter 压测，大部分都是测试的工作。但在测出服务性能问题后，需要研发人员及时找到性能瓶颈并优化。至于线上问题定位与性能调优，是程序员在成长路上的宝贵经验，需要一定的积累。遇到的问题越多，解决问题的能力就会提升得越快。Netty 的应用场景非常广，不只是做 RPC，它在 IM 即时通信、物联网及游戏开发中都被大量使用。

已经到了本书的末尾了，鉴于此，写一段简短的总结：本书详细地介绍了 Netty 的核心源码；采用 Netty 完成了一套高性能的分布式 RPC；对 Netty 时间轮的灵活运用、Netty 各种特性原理进行了详细的剖析，如背压、编码和解码、零拷贝等。有了这些知识，读者对 Netty 的应用会变得更加灵活、自信。